T0226253

Communications
in Computer and Information Science 811

Commenced Publication in 2007
Founding and Former Series Editors:
Alfredo Cuzzocrea, Xiaoyong Du, Orhun Kara, Ting Liu, Dominik Ślęzak,
and Xiaokang Yang

More information about this series at http://www.springer.com/series/7899

Samir Mbarki · Mohammed Mourchid
Max Silberztein (Eds.)

Formalizing Natural Languages with NooJ and Its Natural Language Processing Applications

11th International Conference, NooJ 2017
Kenitra and Rabat, Morocco, May 18–20, 2017
Revised Selected Papers

 Springer

Editors
Samir Mbarki
Université Ibn Tofail de Kénitra
Kenitra
Morocco

Max Silberztein
Université de Franche-Comté
Besançon
France

Mohammed Mourchid
Université Ibn Tofail de Kénitra
Kenitra
Morocco

ISSN 1865-0929   ISSN 1865-0937 (electronic)
Communications in Computer and Information Science
ISBN 978-3-319-73419-4   ISBN 978-3-319-73420-0 (eBook)
https://doi.org/10.1007/978-3-319-73420-0

Library of Congress Control Number: 2017962897

Printed on acid-free paper

This Springer imprint is published by Springer Nature
The registered company is Springer International Publishing AG
The registered company address is: Gewerbestrasse 11, 6330 Cham, Switzerland

Editors' Preface

NooJ is a linguistic development environment that provides tools for linguists to construct linguistic resources that formalize a large gamut of linguistic phenomena: typography, orthography, lexicons for simple words, multiword units and discontinuous expressions, inflectional, derivational and agglutinative morphology, local, structural dependency and transformational syntax, and semantics. For each elementary linguistic phenomenon to be described, NooJ proposes a set of computational formalisms, the power of which ranges from very efficient finite-state automata to very powerful Turing machines as well as a rich toolbox that allows linguists to construct, maintain, test, debug, accumulate, and reuse linguistic resources. This makes NooJ's approach different from most other computational linguistic tools that typically offer a unique formalism to their users and are not compatible with each other.

NooJ provides parsers that can apply any set of linguistic resources to any corpus of texts, to extract examples or counter-examples, annotate matching sequences, perform statistical analyses, and so on. Because NooJ's linguistic resources are neutral, they can also be used by NooJ's generators to produce texts. By combining NooJ's parsers and generators, one can construct sophisticated NLP (natural language processing) applications such as MT (machine translation) systems, paraphrases generators, etc.

Since its first release in 2002, NooJ has been enhanced with new features every year. Linguists, researchers in social sciences, and more generally all professionals who analyze texts have contributed to its development and participated in the annual NooJ conference. In 2013, a new version of NooJ was released, based on the JAVA technology and available to all as an open source GPL project. Moreover, several private companies are now using NooJ to construct business applications in several domains, from business intelligence to opinion analysis. To date, there are NooJ modules available for over 50 languages; more than 3,000 copies of NooJ are being downloaded each year.

The present volume contains 20 articles selected from the papers and posters presented at the International NooJ 2017 conference at the Ibn Tofail University and the ENSIAS, in Morocco. These articles are organized in four parts: "Vocabulary and Morphology" containing five articles; "Syntactic Analysis" containing six articles, "Natural Language Processing Applications" containing seven articles, and "NooJ's Future" containing two articles.

The articles in the first part involve the construction of electronic dictionaries and the description of morphological phenomena, as well as a bilingual comparison of verb tenses information that can be used by MT software:

– Masako Watabe's article "A NooJ Dictionary for the Rromani Language: Toward a NooJ-Relevant Sorting of Morphosyntactic Tags" aims at defining a set of morphosyntactic tags that can be used to describe the properties of substantive in Rromani's four dialects.

- Maximiliano Duran's article "Morphological Grammars to Generate and Annotate Verb Derivation in Quechua" presents the formalization of bi- and tri-suffixed agglutination of verbs in Quechua.
- Cristina Mota, Lucília Chacoto, and Anabela Barreiro's article "Integrating the Lexicon-Grammar of Predicate Nouns with Support Verb *fazer* into Port4NooJ" describes the formalization of 3,000 predicate nouns in Portuguese, and its application in an automatic paraphrase generator.
- Rafik Kassmi, Mohammed Mourchid, Abdelaziz Mouloudi, and Samir Mbarki's article "Processing Agglutination with a Morpho-Syntactic Graph in NooJ" shows how agglutinative morphological grammars (rather than inflectional ones) could be used to formalize Arabic tenses.
- Ilham Blanchete, Mohammed Mourchid, Samir Mbarki, and Abdelaziz Mouloudi's article "Formalizing Arabic Inflectional and Derivational Verbs Based on Root and Pattern Approach Using NooJ Platform" describes a system of dictionary to formalize the Arabic vocabulary, based on roots rather than on lemmas.

The articles in the second part describe the construction of sophisticated syntactic grammars and the use of such grammars to help students learn Italian or Spanish as a second language:

- Nadia Ghezaiel Hammouda and Kais Haddar's article "Arabic NooJ Parser: Nominal Sentence Case" presents a formalization of Arabic nominal sentences and its implementation using NooJ grammars.
- Said Bourahma, Mohammed Mourchid, Samir Mbarki, and Abdelaziz Mouloudi's article "The Parsing of Simple Arabic Verbal Sentences Using NooJ Platform" presents a parser for simple Arabic verbal sentences that uses a dependency grammar to produce parse trees.
- Krešimir Šojat, Kristina Kocijan, and Božo Bekavac's article "Identification of Croatian Light Verb Constructions with NooJ" presents a set of linguistic resources used to formalize light verb constructions in Croatian.
- Maddalena della Volpe, Annibale Elia, and Francesca Esposito's article "Semantic Predicates in the Business Language" presents a set of syntactic grammars that recognize simple sentences in the language used for business, and produce the corresponding semantic predicates.
- Ignazio Mauro Mirto and Emanuele Cipolla's article "Invalid Syntax: NooJ Assisted Automatic Detection of Errors in Auxiliaries and Past Participles in Italian" presents a formalization of compound tenses that can be used to help Italian learners select auxiliary verbs and apply past participle agreements correctly.
- Andrea Rodrigo, Silvia Reyes, and Rodolfo Bonino's article "Some Aspects Concerning the Automatic Treatment of Adjectives and Adverbs in Spanish: A Pedagogical Application of the NooJ Platform" presents a formalization of Spanish Adjective phrases and its pedagogical applications to help Spanish learners.

The articles in the third part involve the construction of software applications capable of parsing and extracting meaningful information:

- Azeddin Rhazi and Ali Boulaalam's article "Corpus-Based Extraction and Translation of Arabic Multi-Words Expressions (MWEs)" presents a hybrid system capable of extracting Arabic multi-word expressions automatically from bilingual corpora.

- Hajer Cheikhrouhou's article "The Automatic Translation of French Verbal Tenses to Arabic Using the Platform NooJ" shows the differences between the Arabic and the French verbs tense systems, and proposes a bilingual set of linguistic resources to translate automatically conjugated French verbs of communication and movement to Arabic.
- Tong Yang's article "Automatic Extraction of the Phraseology Through NooJ" presents a system capable of automatically recognizing and extracting multiword units and semi-frozen expressions from a corpus of French texts in the culinary domain.
- Yuraś Hiecevič, Alena Kryvaltsevich, Nastassia Kazloŭskaja, Anastasija Drahun, Jaŭhienija Zianoŭka, and Aliaksandr Ščarbakoŭ's article "Sentiment Analysis Algorithms for the Belarusian NooJ Module in the Touristic Sphere" presents a software application and its linguistic resources capable of performing automatic sentiment analyses of touristic texts.
- Imen Ennasri, Sondes Dardour, Héla Fehri, and Kais Haddar's article "Question–Response System Using the NooJ Linguistic Platform" presents a question-answering application in the medical domain capable of parsing users' questions in Arabic, which retrieves the potential answers in two corpora of texts: one in Arabic and one in English.
- Mario Monteleone, Raffaele Guarasci, and Alessandro Maisto's article "NooJ Morphological Grammars for Stenotype Writing" presents a system that automatically detects and correct typos in stenotype writing.
- Carmela Scoppetta, Anastasia Alfieri, Flavio Merenda, Sonia Lay, Annalisa Colasanto, and Raffaele Manna's article "From Language to Social Perception of Immigration" presents a system that automatically analyzes a corpus of journalistic texts and a corpus of comments on these texts, on the topic of immigration in Italy.

The articles in the last part involve the development of two companion applications for NooJ: a Web-based graphical interface as well as an industrial-strong linguistic engine:

- Zineb Gotti, Samir Mbarki, Sara Gotti, and Naziha Laaz's article "Nooj Graphical User Interfaces Modernization" presents a theoretical approach to software modernization, and applies it to modernize NooJ's graphical user interface.
- Max Silberztein's article "A New Linguistic Engine for NooJ: Parsing Context-Sensitive Grammars with Finite-State Machines" presents a set of algorithms that can be used to apply linguistic resources developed with NooJ in a very efficient way.

This volume should be of interest to all users of the NooJ software because it presents the latest development of its linguistic resources as well as its future enhancements.

Linguists as well as computational linguists who work on Arabic, Belarusian, Croatian, French, Italian, Portuguese, Quechua, Rromani, or Spanish will find advanced, up-to-the-minute linguistic studies for these languages in this volume.

We think that the reader will appreciate the importance of this volume, both for the intrinsic value of each linguistic formalization and the underlying methodology as well as for the potential for developing NLP applications along with linguistic-based corpus processors in the social sciences.

December 2017

Samir Mbarki
Mohammed Mourchid
Max Silberztein

Organization

Program Committee

Max Silberztein (Program Chair)	Université de Franche-Comté, France
Xavier Blanco	Autonomous University of Barcelona, Spain
Mohammed El Hannach	Sidi Mohammed Ben Abdellah University, Morocco
Mohammed Essaidi	ENSIAS, Morocco
Héla Fehri	University of Gabes, Tunisia
Yuras Hetsevich	United Institute of Informatics Problems, Belarus
Kristina Kocijan	University of Zagreb, Croatia
Svetla Koeva	University of Sofia, Bulgaria
Peter Machonis	Florida International University, USA
Samir Mbarki	Ibn Tofail University, Morocco
Slim Mesfar	University of Manouba, Tunisia
Mohammed Mourchid	Ibn Tofail University, Morocco
Mario Monteleone	University of Salerno, Italy
Johanna Monti	University of Sassari, Italy
Fadoua Ataa Allah	Institut Royal de la Culture AMazighe, Morocco
Jan Radimský	University of South Bohemia, Czech Republic
Azeddine Rhazi	Cadi Ayyad University, Morocco
François Trouilleux	Université Blaise-Pascal, France

Contents

Natural Language Processing Applications

NooJ's Future

List of Contributors

Anastasia Alfieri Dipartimento di Scienze Sociali Politiche e della Comunicazione, Università di Salerno, Fisciano, Italy

Anabela Barreiro L2F/INESC-ID, Lisbon, Portugal

Božo Bekavac Department of Linguistics, Faculty of Humanities and Social Sciences, University of Zagreb, Zagreb, Croatia

Ilham Blanchete MIC Research Team, Laboratory MISC, IbnTofail University, Kenitra, Morocco

Rodolfo Bonino IES N° 28 "Olga Cossettini", Rosario, Argentina

Ali Boulaalam FLSF, Med Ben Abdellah University, Fes, Morocco

Said Bourahma MISC Laboratory, Faculty of Science, Ibn Tofail University, Kenitra, Morocco

Lucília Chacoto Universidade do Algarve, Faro, Portugal; CLUL, Lisbon, Portugal

Hajer Cheikhrouhou University of Sfax, LLTA, Sfax, Tunisia

Emanuele Cipolla Consiglio Nazionale delle Ricerche, Palermo, Italy

Annalisa Colasanto Dipartimento di Scienze Sociali Politiche e della Comunicazione, Università di Salerno, Fisciano, Italy

Sondes Dardour MIRACL Laboratory, University of Sfax, Sfax, Tunisia

Anastasija Drahun The United Institute of Informatics Problems, National Academy of Sciences of Belarus, Minsk, Belarus

Maximiliano Duran Université de Franche-Comté, Besançon, France

Annibale Elia Department of Political, Social and Communication Sciences, University of Salerno, Fisciano, Italy

Imen Ennasri MIRACL Laboratory, University of Sfax, Sfax, Tunisia

Francesca Esposito Department of Political, Social and Communication Sciences, University of Salerno, Fisciano, Italy

Héla Fehri MIRACL Laboratory, University of Sfax, Sfax, Tunisia

Sara Gotti Faculty of Science, Ibn Tofail University, Kenitra, Morocco

Zineb Gotti Faculty of Science, Ibn Tofail University, Kenitra, Morocco

Raffaele Guarasci Dipartimento di Scienze Politiche, Sociali e della Comunicazione, Università degli Studi di Salerno, Salerno, Italy

Kais Haddar MIRACL Laboratory, Faculty of Sciences of Sfax, University of Sfax, Sfax, Tunisia

Nadia Ghezaiel Hammouda Miracl Laboratory, Higher Institute of Computer and Communication Technologies of Hammam Sousse, Sousse, Tunisia

Yuras Hetsevich The United Institute of Informatics Problems, National Academy of Sciences of Belarus, Minsk, Belarus

Rafik Kassmi MISC, Ibn Tofail University, Kénitra, Morocco

Nastassia Kazloŭskaja The United Institute of Informatics Problems, National Academy of Sciences of Belarus, Minsk, Belarus

Kristina Kocijan Department of Information and Communication Sciences, Faculty of Humanities and Social Sciences, University of Zagreb, Zagreb, Croatia

Alena Kryvaltsevich The United Institute of Informatics Problems, National Academy of Sciences of Belarus, Minsk, Belarus

Naziha Laaz Faculty of Science, Ibn Tofail University, Kenitra, Morocco

Sonia Lay Dipartimento di Scienze Sociali Politiche e della Comunicazione, Università di Salerno, Fisciano, Italy

Alessandro Maisto Dipartimento di Scienze Politiche, Sociali e della Comunicazione, Università degli Studi di Salerno, Salerno, Italy

Raffaele Manna Dipartimento di Scienze Sociali Politiche e della Comunicazione, Università di Salerno, Fisciano, Italy

Samir Mbarki MIC Research Team, MISC Laboratory, Faculty of Science, Ibn Tofail University, Kenitra, Morocco

Flavio Merenda Dipartimento di Scienze Sociali Politiche e della Comunicazione, Università di Salerno, Fisciano, Italy

Ignazio Mauro Mirto Università di Palermo, Palermo, Italy

Mario Monteleone Dipartimento di Scienze Politiche, Sociali e della Comunicazione, Università degli Studi di Salerno, Salerno, Italy

Cristina Mota L2F/INESC-ID, Lisbon, Portugal

Abdelaziz Mouloudi MIC Research Team, MISC Laboratory, Faculty of Science, Ibn Tofail University, Kenitra, Morocco

Mohammed Mourchid MIC Research Team, MISC Laboratory, Faculty of Science, Ibn Tofail University, Kenitra, Morocco

Silvia Reyes Facultad de Humanidades y Artes, Universidad Nacional de Rosario, Rosario, Argentina

Azeddin Rhazi FLSH, Qadi Ayyad University, Marrakech, Morocco

Andrea Rodrigo Facultad de Humanidades y Artes, Universidad Nacional de Rosario, Rosario, Argentina

Aliaksandr Ščarbakoŭ The Belarusian State University of Informatics and Radio-electronics, Minsk, Belarus

Carmela Scoppetta Dipartimento di Scienze Sociali Politiche e della Comunicazione, Università di Salerno, Fisciano, Italy

Krešimir Šojat Department of Linguistics, Faculty of Humanities and Social Sciences, University of Zagreb, Zagreb, Croatia

Max Silberztein Université de Franche-Comté, Besançon, France

Maddalena della Volpe Department of Business, Management and Innovation System, University of Salerno, Fisciano, Italy

Masako Watabe Paris-Sorbonne University, Paris, France

Tong Yang DILTEC (Didactiques des langues, des textes et des cultures), Université Sorbonne Nouvelle (Paris 3), Paris, France

Jaŭhienija Zianoŭka The United Institute of Informatics Problems, National Academy of Sciences of Belarus, Minsk, Belarus

Vocabulary and Morphology

A NooJ Dictionary for the Rromani Language: Toward a NooJ-Relevant Sorting of Morphosyntactic Tags

Masako Watabe(✉)

Paris-Sorbonne University, Paris, France
masako.watabe@paris-sorbonne.fr

Abstract. This paper aims at presenting how to elaborate a relevant sorting of morphosyntactic tags to be used in the NooJ dictionary for Rromani language through three topics: dialectal issues, treatment of postpositions and countableness of substantives. This module encompasses all four dialects of Rromani, the isoglosses of which are basically no longer geographical. We have thus defined each of the four dialects through a combination of two tags corresponding to specific isoglosses. For instance, the so-called O-bi dialect (i.e. O-superdialect with no mutation of alveolar affricates) is labelled as "rro + rrbi" in NooJ. Then, on typological grounds, it was decided to treat the Rromani postpositions as agglutinative, non-inflectional, morphemes. Rromani postpositions are appended to substantives in the oblique case and in some cases cumulative (as in Modern Indic). In addition, the postposition of possession may be inflected in gender, number and case as an adjective (**-qo, -qi, -qe** *of* as basic forms, with variants). Accordingly, no less than some 250 potential forms are to be encountered for postpositions, covering all basic dialectal variants. However, they may all be rendered, by a much more economical system, appropriate to both Rromani grammar and computational analysis. Moreover, we investigated the system of countableness in Rromani nouns when relevant.

Keywords: NooJ · Rromani language · Morphosyntax · Rromani dialectology
Postpositions · Countable

1 Introduction

1.1 Rromani Language (i Rromani Ćhib)

Rromani is the mother tongue of the Rromani people (ca. 15 M persons), scattered throughout all European and American countries. It is spoken on a daily basis by about five and a half million Rroms.

A language spoken by an Indo-Aryan people[1], Rromani belongs to the Indo-Aryan subfamily, and is therefore closely related to present-day northern Indian languages, not only in vocabulary but also in grammar (especially the nominal group system) [1]. In fact, Rromani is one of the Modern Indic languages most closely related to Sanskrit.

---

[1] Proto-Rromani people were deported by Mahmood Sultan in 1018 from Kanauji in the middle valley of the Ganges.

© Springer International Publishing AG 2018
S. Mbarki et al. (Eds.): NooJ 2017, CCIS 811, pp. 3–15, 2018.
https://doi.org/10.1007/978-3-319-73420-0_1

1.2 NooJ Module for the Rromani Language

An innovative feature of the Rromani module is that it aims at covering several varieties of the language, namely all four dialects[2] of Rromani; its structure is polylectal. In other words, one single module for Rromani is designed to formalize the entire Rromani language, giving all four dialects equality standing.

The NooJ module for Rromani is rich in morphology; the main paradigms of nouns, adjectives and verbs are programed, including exceptions and interdialectal variants in order to formalize the inflectional morphology and the dialectology. However, the number of entries in the dictionary is still small. To complete the NooJ dictionary for Rromani, we would import an editorial dictionary [2], which contains about 10,000 entries and covers all main dialects. Then, how is it possible to build a coherent system for morphosyntactic tags at this stage?

2 Rromani Dialects

2.1 Superdialects: O and E

Two types of isoglosses that are non-areal and crossed define the division into four dialects of the Rromani language.

The first isogloss is the opposition "o *vs.* e" forming the two superdialects:

- O-superdialect,
- E-superdialect.

This opposition is systematically marked on the endings of the verbs (1SG.PST. IND), the copula (1SG.PRS.IND) and the definite article (PL.DRT) (Table 1).

Table 1. Opposition "o *vs.* e"

O-superdialect "rro"	E-superdialect "rre"	
phirdom, phirdŏm, phirdŭm	**phirdem**	(*I*) *walked*
sinom, som, sinŏm, hom, hium	**sem**	(*I*) *am*
o Rroma	**e Rroma**	*the Rroms*

In the Rromani module, two tags "rro" and "rre" represent the two superdialects: O and E.

2.2 Phonetic Mutation

The criterion of the second isogloss is a phonetic mutation that forms the two dialectal subgroups:

---

[2] The four dialects of the Rromani are respectively called O-bi, O-mu, E-bi and E-mu (see chap. 2).

- without phonetic mutation,
- with phonetic mutation.

The two consonants "ćh" and "ʒ" are originally alveolar affricates [tʃʰ] and [dʒ], but they can be transformed into alveolo-palatal fricatives [ɕ] and [ʑ]. This mutation concerns only the phonetic realization, not the spelling. Therefore, these variants are not treated in NooJ, which addresses only the written form of the text (Table 2).

Table 2. Opposition "alveolar affricates *vs.* alveolo-palatal fricatives"

Without mutation "rrbi"	With mutation "rrmu"	
ćhavo [tʃʰavo]	**ćhavo [ɕavo]**	*Rromani boy, son*
ʒukel [dʒukel]	**ʒukel [ʑukel]**	*dog*

These subgroups are represented by two tags "rrbi" and "rrmu". In Rromani, **bi** means *without*, and "mu" is the beginning of the noun **mutàcia** *mutation*.

2.3   Double Tags for a Single Dialect

The entire Rromani language is divided into two superdialects (O and E), and then each of the superdialects is divided into two subgroups (without or with phonetic mutation). The four dialects of the Rromani language have been formed because of the two crossed isoglosses. These four dialects are named "O-bi" (O-superdialect without mutation), "O-mu" (O-superdialect with mutation), "E-bi" (E-superdialect without mutation) and "E-mu" (E-superdialect with mutation) (Table 3).

Table 3. Four dialects of Rromani

	Without mutation	With mutation
O-Superdialect	O-bi "rro + rrbi"	O-mu "rro + rrmu"
E-Superdialect	E-bi "rre + rrbi"	E-mu "rre + rrmu"

A single tag cannot represent a Rromani dialect. Each dialect is defined by two characteristics; one is common with another dialect belonging to the same superdialect, and the other is common with another dialect belonging to the same phonetic subgroup. For example, O-bi and O-mu dialects have an identical verbal morphology **phirdom** (*I*) *walked*. On the other hand, O-bi and E-bi dialects have an identical phonetic realization of the consonant "ʒ" [dʒ] despite the difference of superdialects. To reconstitute this crucial phenomenon of the Rromani dialectology, each dialect should be defined by a combination of two tags: "rro + rrbi" for the O-bi dialect, "rro + rrmu" for the O-mu, "rre + rrbi" for the E-bi, and "rre + rrmu" for the E-mu.

For example, there are three diasynonyms of the negative adverb in Rromani: **na** *not* in the O-superdialect, **ni** *not* in the E-bi dialect, **ći** *not* in the E-mu dialect. For instance, these diasynonyms are programed as below in the NooJ dictionary for Rromani.

```
na,ADV+neg+rro+SYN="ni,ći"+EN="not"
ni,ADV+neg+rre+rrbi+SYN="na,ći"+EN="not"
ći,ADV+neg+rre+rrmu+SYN="na,ni"+EN="not"
```

All these entry words **na**, **ni**, **ći** are negative adverbs (ADV + neg) and mean *not* in English (EN). However, each lemma belongs to a different dialectal group: **na** to the O-superdialect (rro), **ni** to the E-bi dialect (rre + rrbi), and **ći** to the E-mu dialect (rre + rrmu). Then, diasynonyms are preceded by "SYN" meaning (dia)**syn**onym. For example, the lemma **na** has two diasynonyms: **ni** and **ći**.

3   Postpositions

In Rromani, there are four invariable postpositions and one variable postposition:

- **-qe**   *for*   dative,
- **-ça**   *with*   instrumental,
- **-θar**   *from*   ablative,
- **-θe**   *at*   locative,
- **-qo/-qi/-qe** *of*   possessive.

Each postposition is appended to a noun in the oblique case to give a functional value[3] (e.g. ablative). The postposition of possession agrees with gender (masculine, feminine), number (singular, plural) and morphological case (direct, oblique) of the possessed substantive. The possessive postposition has short and long forms, among the long forms there are several dialectal variants. As a result, each noun could be associated with a set of some 250 forms of the possessive postposition.

3.1   Inflection or Agglutination?

In the initial module for Rromani, the postpositions have been programed in the NooJ inflectional morphology (.nof file). This grammar is applied to a NooJ lexical dictionary (.dic file) for Rromani, and then the NooJ system generates automatically more than 250 "forms"[4] to each nominal entry. The "inflected" forms are stored in the inflectional NooJ dictionary (-flx.dic file). However, is such a long list of "forms" with postpositions indispensable? Yet, the postpositions are common to all nouns except the restriction of use according to dialects[5].

---

[3] In Rromani grammar, two levels of cases should be distinguished sharply: the two morphological cases (direct, oblique) expressed by an inflectional ending, and several functional cases (e.g. ablative) expressed, either by a postposition appended to a noun in the oblique case, or by a preposition preceding a noun in the direct case. Prepositional and postpositional phrases could be often equivalents (e.g. **e raklesθar** *from the boy* with a postposition **-θar** *from* vs. **katar o raklo** *from the boy* with a preposition **katar** *from*).

[4] For example, a noun **raklo** *boy* generates 257 "forms" in total: seven forms without postposition, 10 forms with invariable postpositions and 240 forms with a variable postposition.

[5] For example, long forms of the postposition **-qo** *of* are used only in the O-bi dialect.

From a pedagogical point of view, the annotation in the initial module was not entirely appropriate. The inflectional information of noun and postposition was confused (Fig. 1). The module users cannot know the position of the separation between the noun and the postposition in the postpositional phrase. Besides, it is not possible to know which information is carried by the noun, and which information is carried by the postposition. In fact, the "form" **raklesqi** *of boy* consists of two elements: **rakles** *boy*, a masculine human noun (N + hum + m) **raklo** *boy* in the singular oblique case (sg + ob), and **-qi** *of*, the possessive postposition (poss) **-qo** *of* in the feminine singular direct case (Df + Dsg + Ddr[6]).

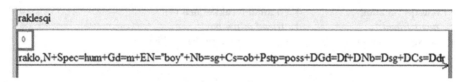

Fig. 1. Annotation of **raklesqi** *of boy* in the initial module

From a linguistic point of view, it would be adequate to treat the Rromani postpositions as agglutination. True, the possessive postposition inflects, but the postposition itself is not an ending.

3.2   Postpositional Phrases in Rromani

In NooJ, there are two types of the morphology: on the one hand inflection and derivation (.nof file), and on the other hand productive morphology (.nom file). In the newest Rromani module, postpositions have been formalized as agglutinative morphemes by a productive morphology.

O ruv si e raklesqo (1) **amal.**   *The wolf is the boy's friend.*
Perel jekh asvin e raklesqe (2) **jakhaθar** (3).   *A tear falls down from the boy's eye.*

In the examples above, there are three occurrences of postpositional phrases numbered from (1) to (3). Each occurrence consists of two ALUs (Atomic Linguistic Units)[7], equivalent to, in this context, a noun and a postposition agglutinated each other.

- **raklesqo** *boy's* (1): the first ALU **rakles-** is a masculine noun **raklo** *boy* in the singular oblique case, and the second ALU **-qo** is the possessive postposition **-qo** *of* in the masculine singular direct case according to the possessed noun **amal** *(male) friend*,
- **raklesqe** *boy's* (2): the first ALU **rakles-** is the same as in **raklesqo** (1), and the second ALU **-qe** is the possessive postposition **-qo** *of* in the feminine singular oblique case according to the possessed noun **jakha-** *eye*,

---

[6] The capital "D" precedes the inflectional or semantic information of determinee (e.g. possessed substantive) in the Rromani module. For example, "Dsg" means the possessed substantive is in the singular case.

[7] In NooJ, words, lexemes and morphemes could be considered as ALUs. [3].

- **jakhaθar** *from eye* (3): the first ALU **jakha-** is a feminine noun **jakh** *eye* in the singular oblique case, and the second ALU **-θar** *from* is the ablative postposition.

In general, the possessive postposition in Rromani is placed before its possessed substantive, yet this order could be reversed providing that the possessive postposition will be rendered a long form (4). Then, if one of the invariable postpositions is appended to a possessed substantive, the same postposition should be appended to the possessive postposition (5) too. This addition is for a grammatical reason and any functional value (e.g. ablative) would not be added, except emphasizing the possessive.

• Neutral	• Emphatic
e raklesqo (1) **amal**	=> **o amal e raklesqoro** (4)
the boy's friend	*the friend of the boy*
e raklesqe (2) **jakhaθar**	=> **e jakhaθar e raklesqerăθar** (5)
from the boy's eye	*from the eye of the boy*

Both two postpositional phrases **raklesqo** (1) *boy's* and **raklesqoro** (4) *of boy* consist of two ALUs. They have identical values at the lexical, semantic and morphological level. The only differences are the length of the possessive postposition (i.e. **-qo** *of* vs. **-qoro** *of*) and the emphatic feature (i.e. **-qo** *of* neutral vs. **-qoro** *of* emphatic).

The number of ALUs is different (i.e. two or three) between other two postpositional phrases **raklesqe** (2) *boy's* and **raklesqerăθar** (5) *of boy*. However, they do not have any difference of lexical, semantic and morphological value either (except emphasis).

- **raklesqerăθar** (5) *of boy*: the first ALU **rakles-** is the same as in **raklesqo** (1) and **raklesqe** (2) above, the second ALU **-qeră-** is the possessive postposition **-qo** *of* in the feminine singular oblique case in the nominal inflection according to the possessed noun **jakha-** *eye*, and the third ALU **-θar** *from* is the ablative postposition but without its functional value (i.e. ablative).

How could these agglutinative ALUs be reconstituted in NooJ? We will see a productive morphology for postpositional phrases in Rromani.

3.3   Productive Morphology for the Rromani

In the productive morphology for Rromani below (Fig. 2), each line corresponds to a postpositional phrase including one or two postpositions. Each line begins with a common variable "Subs" encompassing a sequence of letters. The command "<L>" represents a letter, and a loop over the node indicates the possibility of unlimited repetition. For instance, "<L>" with a loop means a sequence of any number of letters. However, the variable "Subs" (i.e. sequence of letters) is restricted by a constraint ":N[8] + ob" meaning any noun in the oblique case. For example, **rakles** *boy* and **jakha-**[9]

---

[8] A colon means inclusiveness. For example, ":N" includes any noun in any inflected form.

[9] Inanimate nouns in the oblique case do not exist without postposition in Rromani.

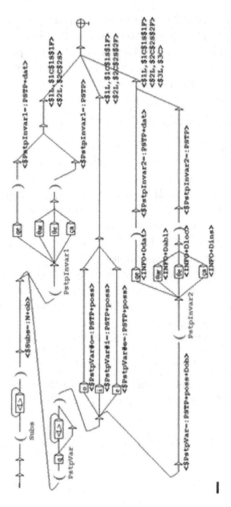

Fig. 2. A grammar for postpositional phrases of the Rromani

eye are nouns in the oblique case, so they match the variable "Subs". On the other hand, **ruv** *wolf* in the direct case does not match.

Then, there are two possibilities as second variable: "PstpInvar1 (invariable postposition 1)" or "PstpVar (variable postposition)". Besides, the third variable "PstpInvar 2 (invariable postposition 2)" could be appended to "PstpVar". Finally, each line arrives at the end.

If the postpositions were treated in the inflectional grammar, it would be enough to have a NooJ lexical dictionary and a NooJ inflectional grammar to provide an annotation. However, the information of all ALUs constituting a postpositional phrase will be mixed in the annotation (Fig. 1).

With a productive morphology, only inflected forms[10], *stricto sensu*, will be required in order that NooJ recognizes both substantives without or with postposition, and it no longer needs a voluminous dictionary including postpositional phrases. Thus, NooJ dictionary for Rromani has been lightened (e.g. 257 forms to seven about a noun **raklo** *boy*), without losing analytic capacities.

3.4   Two ALUs with an Invariable Postposition

When a productive morphology is applied to a text, NooJ will analyze each phrase to identify possible combinations of ALUs according to the constraints. For example, an inflected form **rakles** *boy* will not be recognized by the grammar for postpositional phrases[11], because it is not followed by any postposition, even if it corresponds to the constraint ":N + ob". On the other hand, a postpositional phrase **raklesθar** *from boy* should be recognized, because **rakles-** *boy* corresponding to the variable "Subs" (i.e. a noun in the oblique case) is followed by **-θar** *from* corresponding to a second variable "PstpInvar1" (i.e. an invariable postposition).

The variable "PstpInvar1" encompasses four invariable postpositions (**-qe** *for*, **-θar** *from*, **-θe** *at*, **-ça** *with*); on the other hand, there is an additional constraint "dat" for the dative postposition **-qe** *for*. It is to distinguish it from the possessive postposition **-qo** *of* having an inflected form **-qe** *of* as well.

At the end of this sequence, there is a set of annotation orders.

```
<$1L,$1C$1S$1F><$2L,$2C$2S>
```
[12]

These orders are separated in two parts; each of which corresponds to an ALU. As a result, each ALU would be annotated separately (Fig. 3).

Fig. 3.   Annotation of **raklesθar** *from boy*

NooJ recognizes that the postpositional phrase **raklesθar** *from boy* matches perfectly the sequence {"Subs" + "PstpInvar1"}. Then, NooJ provides its annotation through the corresponding annotation orders.

---

[10] In general, a noun inflects in four forms; a masculine noun **raklo** *boy* inflects in: **raklo** *boy* (sg + dr), **rakles** *boy* (sg + ob), **rakle** *boys* (pl + dr), **raklen** *boys* (pl + ob).

[11] However, a NooJ inflectional dictionary would recognize it.

[12] Each constraint (and its variable) is numbered from left to right ($1 being the first constraint), and the various fields of the lexicon are named "L" (corresponding Lemma), "C" (morphosyntactic Category), "S" (Syntactic or semantic features) and "F" (inFlectional information). For instance, "$1L" means corresponding lemma of the first constraint. [4].

```
<$1L,$1C$1S$1F>      => raklo,noun + human + boy + masculine + singular + oblique
<$2L,$2C$2S>         => θar,postposition + ablative + from
```

Concerning the first variable ($1), corresponding lemma ($1L) is **raklo**: category ($1C) is noun (N), semantic feature ($1S) is human (hum), and inflectional information ($1F) is masculine singular oblique (m + sg + ob). For the second variable ($2), corresponding lemma ($2L) is **-θar**: category ($2C) is postposition (PSTP), semantic feature ($2S) is ablative (abl). There is no "$2F" (i.e. inflectional information of the second variable), because -θar is an invariable postposition.

In this postpositional phrase, which consists of two ALUs, the annotation is clearly separated in two parts. Thus, there is no risk to confuse the information of one ALU with another.

3.5   Two ALUs with a Variable Postposition

The variable "Subs" (i.e. a noun in the oblique case) is followed by not only "PstpInvar1", but also "PstpVar" corresponding to the postposition of possession that is morphologically variable.

The basic form of the possessive postposition is **-qo** *of*. There are four dialectal variants in long forms: **-qoro**, **-qro**, **-qëro**, **-qero** *of*. Whatever variant and inflected form, if the possessive postposition does not have any other postposition appended, it will end with one of the three vowels: "-o", "-i" or "-e" according to the gender, number and case of the possessed substantive. The inflection with three final vowels "-o", "-i" and "-e" corresponds to an adjectival paradigm, named **buxlo** *large*[13].

- Adjectival inflection

Dm + Dsg + Ddr:	**e raklesqo kalo bal**	*the boy's black hair*
Df + Dsg + Ddr:	**e raklesqi kali jakh**	*the boy's black eye*
Dm + Dpl + Ddr:	**e raklesqe kale bala**	*the boy's black hairs*
Df + Dpl + Ddr:	**e raklesqe kale jakha**	*the boy's black eyes*
Df + Dsg + Dob:	**e raklesqe kale jakhaθar**	*from the boy's black eye*

In the grammar above (Fig. 2), there are three vowels "o", "i" and "e" each of which is put in a node over the constraint "#o", "#i" or "#e". This means the variable "PstpVar" should end with one of these vowels. The variable "PstpVar" encompasses the initial letter "q-", and either, a sequence of letters or an empty string, then the final letter "-o", "-i" or "-e". This formalizes all variants of the possessive postposition: **-qo** *of* (no letter between "q-" and "-o"), **-qro** *of* (one letter between "q-" and "-o"), **-qoro**, **-qëro**, **-qero** *of* (two letters between "q-" and "-o"), and their inflected forms. At the end of this sequence, there is a very similar set of annotation orders as we have seen in another type of postpositional phrase {"Subs" + "PstpInvar1"}.

---

[13] The paradigm **buxlo** *large* covers all adjectives, which are vocalic (i.e. ending "-o" in the basic form) and oxytonic (e.g. **buxlo** *large*, **kalo** *black*).

Fig. 4. Annotation of **raklesqoro** *of boy*

```
<$1L,$1C$1S$1F><$2L,$2C$2S$2F>
```

The only difference is the addition of "$2F" (i.e. inflectional information of the second variable) as the possessive postposition inflects.

NooJ recognizes that the phrase **raklesqoro** *of boy* (4) matches perfectly the sequence {"Subs" + "PstpVar"}. Then, NooJ provides its annotation (Fig. 4) through the corresponding annotation orders.

<$1L,$1C$1S$1F> => **raklo**, noun + human + *boy* + masculine + singular + oblique

<$2L,$2C$2S$2F> => **qo**, postposition + possessive + *of* + O-bi dialect[14] + southern   vernacular[15] + masculine + singular + direct (concerning the possessed noun)[16]

The annotation of **raklesqoro** *of boy* is separated in two parts clearly and correctly.

3.6   Three ALUs with a Variable Postposition and an Invariable Postposition

The variable "PstpVar" (i.e. variable postposition) can be followed by another invariable postposition, too. In this case, the possessive postposition inflects according to the nominal pattern.

- Nominal inflection

Dm + Dsg + Dob: **e phralesθar e raklesqeresθar**   from the brother of the boy
Df + Dsg + Dob: **e phenăθar e raklesqerăθar**   from the sister of the boy
Dm + Dpl + Dob: **e phralenθar e raklesqerenθar**   from the brothers of the boy
Df + Dpl + Dob: **e phenĕnθar e raklesqerĕnθar**[17]   from the sisters of the boy

The possessive postposition with a nominal ending can exist only with another invariable postposition appended. The constraints about the final letters "o", "i" and "e" are useful not to identify non-existent forms (e.g. *****raklesqeres**).

---

[14] Remember the combination of two tags "rro + rrbi" represents the O-bi dialect.

[15] The tag "rrs" (as south) represents the vernacular used in the Balkans.

[16] Remember the capital "D" precedes the information of determinee.

[17] These inflected forms of the possessive postposition are used in either the Balkan vernacular or the Carpathian one, both belonging to the O-bi dialect.

The constraints of "PstpInvar2" are identical with "PstpInvar1". On the other hand, there is additional information (e.g. "<INFO + Dabl>"), and it lacks "$3S" in the annotation orders. Why?

Fig. 5. Annotation of **raklesqërăθar** *of boy*

If there is an annotation order "$3S", NooJ will show not only the semantic feature "Dabl"[18], but also the translation *from*. In fact, in the agglutinative phrase including two postpositions (e.g. **raklesqërăθar** *of boy*), the second postposition will lose its semantic value. The reason why "$3S" has been removed, and "<INFO+Dabl>" is added. Thus, we could have a more appropriate annotation, only with "Dabl" but no translation *from* (Fig. 5).

<$1L,$1C$1S$1F>   => **raklo**,
                     noun + human + *boy* + masculine + singular + oblique
<$2L,$2C$2S$2F>   => **qo**,   postposition + possessive + *of* + O-bi   dialect +
                     northern vernacular[19] + feminine + singular + oblique (con-
                     cerning the possessed noun)
<$3L,$3C>   => **θar**, postposition + ablative (concerning the possessed
                     noun)

Through the grammar for postpositional agglutination, most of the postpositional phrases are recognized and annotated successfully. For example, among 250 postpositional phrases with a noun **raklo** *boy*, only one phrase is not recognized, and four other phrases are recognized but their first ALUs are not correctly annotated.

4 Countableness

In Rromani, nouns are divided into two groups according to countableness: countable and uncountable.

---

[18] Remember that "Dabl" means the posssessed noun is in the ablative case, not the possessor.

[19] The tag "rrn" (as **n**orth) represents the vernacular used in Russia and the north of Poland.

Countable nouns have two morphological forms (singular, plural) according to their number (only one or at least two): **jekh**[20] **raklo** *a boy*, **duj rakle** *two boys*; **jekh ruv** *a wolf*, **duj ruva** *two wolves*; **jekh kher** *a house*, **duj khera** *two houses*.

Uncountable nouns are used mainly in the singular form whatever quantity is: **xàca mas** *a little meat*, **but mas** *a lot of meat*. The plural form of uncountable nouns can be used to express the plurality of sorts: **mas** *one kind of meat*, **masa** *mixed meat with different kinds*. On the other hand, the plural form of uncountable nouns can be used to mean very large quantity: **but masa** *abundant meat*.

Some uncountable nouns have only plural forms: **lima** *snot*, **pixa** *eye-rheum*.

Besides, some nouns have a different semantic value in the singular or plural form: **mas** *meat*, **masa** *muscle*; **jekh lovorro**[21] *a small piece of money*, **love** *money*.

```
raklo,N+hum+m+EN="boy"+FLX=ćhavo
ruv,N+ani+m+EN="wolf"+FLX=rrom
mas,N+ina+m+EN="meat"+FLX=kher
lima,N+ina+f+pl+EN="snot"+FLX=lima
love,N+ina+m+pl+EN="money"+FLX=love
lovorro,N+ina+m+EN="piece of money"+FLX=ilo
```

As we can see, some entry words are extracted from the NooJ dictionary above. There is no tag to indicate the countableness. In fact, only the gender (masculine, feminine), the nature of entity (human, animal, inanimate) and, if it is necessary, the number (plural) are indicated as morphosyntactic tags for substantives in the present module. Then, even if the diminutive may be programed as derivational morphology in the Rromani module, and **lovorro** *piece of money* is a diminutive of *lovo (i.e. singular form of **love** *money*), **lovorro** should be an entry word of dictionary because it has a different semantic value.

To build the system of countableness in Rromani nouns for the NooJ module, it would be necessary to study the entire lexicon of the editorial dictionary and to carry out more syntactic studies in depth.

5 Conclusion and Future Perspectives

The productive morphology for postpositional phrases functions mostly well. However, to identify and annotate them faultlessly, there are some obstacles: such as the ambiguity between the dative postposition **-qe** *for* and an inflected form **-qe** of the possessive postposition **-qo** *of*. To remove ambiguities, we should investigate the syntactic studies.

Concerning the dialect issues and uncountable nouns, we need to analyze a more important lexicon through an editorial dictionary to build a complete system.

---

[20] In Rromani, there is no indefinite article. However, the cardinal number **jekh** *one* is used as the singular indefinite article.

[21] On morphological ground, the singular form of **love** *money* is *lovo, yet its diminutive **lovorro** is used as an equivalent.

References

1. Courthiade, M.: The nominal flexion in Rromani. In: Courthiade, M., Grigore, D. (eds.) Professor Gherghe Sarău: a Life Devoted to the Rromani Language. Editura universității din bucurești, Bucharest (2016)
2. Courthiade, M., et al.: Morri angluni rromane ćhibăqi evroputni lavustik. Cigány Ház, Budapest (2009)
3. Silberztein, M.: La formalisation des langues: l'approche de NooJ. ISTE Eds., London (2015)
4. Silberztein, M.: NooJ Manual (2003). www.nooj4nlp.net

Morphological Grammars to Generate and Annotate Verb Derivation in Quechua

Maximiliano Duran[(⊠)]

Université de Franche-Comté, Besançon, France
duran_maximiliano@yahoo.fr

Abstract. In this paper we, study the morphology of Quechua verbs. In order to program their grammars, we first present the formalization of the agglutination laws of verbal suffixes in the form of NooJ morphological grammars. We isolate and study a class of highly productive bi and tri-suffixed agglutinations of interposed suffixes IPS. The derivations obtained, new atomic linguistic con-jugable units ALUc's opens the way to favorably answer the challenging problem of translation of thousands of French verbs into Quechua, a language that contains less than 1 500 simple verbs.

Keywords: Quechua verb morphology · Suffix agglutination
Interposed suffix · NooJ grammars · Quechua verb derivation
Post posed suffixes · Verbalizers

1 Introduction

Quechua is the language of the Incas, an ancient civilization in South America. It's spoken by over 6 million people in Peru, Equator, Bolivia, and Argentina.

Concerning the verbs we notice two important characteristics: All the Quechua verbs are regular verbs, typologically speaking it is a SOV language.

We show in this paper how to generate new Quechua verbs through the derivation of simple verbs. We show also how to annotate them using certain morphological grammars programmed with the help of NooJ (Silberztein 2003, 2010). To reach this objective we should go through the following steps: first, we identify the basic canonical conjugated verbal form and isolate two classes of suffixes, the Inter positional suffixes IPS (Duran 2009, 2013), which will allow us to obtain new verbs and the post-posed suffixes PPS[1] which will present different modalities of a given verb without changing the part of speech. Parker calls it "modal suffix system" (Parker 1969). Then we describe in detail some of these generating paradigms in the NooJ environment (Silberztein 2016). We show some details of the programs that we have built to annotate the inflected verbs, then we describe how we can use this method to obtain the Quechua-French translation of these verbs with the help of automatically generated annotations.

---

[1] Parker (1969) unifies both sets in a single one and call it «modal suffix system». He considers only 16 suffixes.

S. Mbarki et al. (Eds.): NooJ 2017, CCIS 811, pp. 16–28, 2018.
https://doi.org/10.1007/978-3-319-73420-0_2

2 The Basic Canonical Conjugated Form

The main canonical conjugation form for any transitive verb follows the scheme

Subject + V STEM + ENDING

In which the person is marked but the time is not as is shown in the following table (Table 1):

Table 1. The unmarked-time basic conjugation of a verb

PRO QU	PRO EN	Verbal stem	PR ending
Ñuqa	I	Stem	+ NI
qam	You	Stem	+ NKI
pay	He, she	Stem	+ N
ñoqanchik	We	Stem	+ NCHIK
ñoqaiku	We (excl.)	Stem	+ NIKU
qamkuna	You	Stem	+ NKICHIK
paykuna	They	Stem	+ NKU

We will use the symbol PR for the set of these basic endings:
PR = (NI, NKI, N, NCHIK, NIKU, NKICHIK, NKU).

The general, non-future, verbal canonical form is:

V stem + IPS suffix+PR Ending+ PPS suffix

Where IPS symbolizes any interposed suffix and PPS symbolizes any postposed suffix: e.g.:

llamka-ri-ni	I begin to work	(1 IPS before the NI ending)
llamka-ni-raq	I first worked	(1 PPS after the NI ending)

A typical verbal form contains one to two agglutinated IPS preceding the ending and one to two PPS following the ending like in the following examples:

llamka-chi-ni-raq	I first made him work	(1IPS before, and 1 PPS after the NI ending)
llamka-na-nchik-raq-mi	We have to do the work putting everything aside	(1IPS before the NCHIK ending and 2 PPS after the NCHIK ending)

In the last two inflected forms *ni* and NCHIK act as fixed points.

As a matter of fact, a remarkable property during the inflection of any transitive verb is that the PR endings behave as fixed points around which IPS or PPS suffixes may be agglutinated to obtain a verbal form. The following table shows some more examples for the *ni* ending:

> *Miku-**ni*** (I eat)
> *Miku-chka-**ni*** (I am eating)
> *Miku-chka-**ni**-raq* (I am eating before anything else)
> *Miku-chi-chka-**ni**-raq-mi* ((I am eating before anything else indeed)
> *Miku-chi-yku-chka-**ni**-lla-raq-mi* (I am carefully helping him to eat before anything else indeed

Fig. 1. The PR ending acts as a fixed point

The PR ending acts as a fixed point.

For the other endings, we may obtain their corresponding form replacing the NI ending by the one we want to use, leaving unchanged the rest of the form.

Thus if we represent by Fni the NI form: stem + IPS + ni + PPS.

And by Tnki the transformation that replaces the NI ending by the NKI ending. We will have:

> Tnki(Fni)=Fnki, more explicitly
> Tnki(stem+IPS+ni+PPS) = stem+IPS+nki+PPS

Where IPS[2] may be an agglutination of 1 to 4 interposition suffixes and PPS[3] may be an agglutination of 1 to 3 postposition suffixes.

3 Verbal Morphology

3.1 PPS Mono Suffixation

The mono suffixed inflection of a transitive verb including the future conjugation using 1 PPS or 1 FS[4] (future conjugation endings) are obtained applying the following NooJ grammar:

> V_SPP1 = :SPP1_C| :SPP1_V |:SPP1_F;

Where
SPP1_V is the paradigm for the endings ending in a vowel and SPP1_C for those ending in a consonant

---

[2] IPS = (*chi, chka, ikacha, ikachi, ikamu, ikapu, ikari, iku, isi, kacha, kamu, kapu, ku, lla, mpu, mu, naya, pa, paya, pu, raya, ri, rpari, rqu,ru,tamu*).

[3] PPS = (ch, chá, chik, chiki, chu, chu(?), chusina, m, mi, má, man, ña, pas, puni, qa, raq, s, si, taq, yá).

[4] FS = (saq, nki,nqa, sun(nchik), saqku, nkichik, nqaku).

```
SPP1_V= ch |chiki |chu? |chu |chusina |má |m |ña |pas |puni |qa |raq |s |taq |yá;
SPP1_C= chá |chiki |chu? |chu |chusina |má |mi |ña |pas |puni |qa |raq |si |taq |yá;
SPP1_F= SPP1_V_F| SPP1_C_F;

SPP1_V_F=(:F_V)(ch |chiki |chu? |chu |chusina |má |m |ña |pas |puni |qa |raq |s |taq
|yá);
SPP1_C_F=(:F_C)(chá |chiki |chu? |chu |chusina |má |mi |ña |pas |puni |qa |raq |si |taq
|yá);
SPP1_F=:SPP1_V_F | :SPP1_C_F;
F_V = <B>(nki/F+s+2 | nqa/F+s+3 | saqku/F+pex+1 | nqaku/F+p+3);
```

Giving a total of 210 inflected forms containing one postposition suffix as shown in the following Fig. 2.

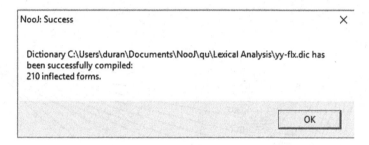

Fig. 2. Inflected forms bearing one interposed suffix

The following is a sample of these inflected forms:

```
rimaytaq,rimay,V+FR="parler"+FLX=V_SPP1+
rimayyá,rimay,V+FR="parler"+FLX=V_SPP1+
rimaychu?,rimay,V+FR="parler"+FLX=V_SPP1+
rimanqakuch,rimay,V+FR="parler"+FLX=V_SPP1+F+p+3
rimanqakuchiki,rimay,V+FR="parler"+FLX=V_SPP1+F+p+3
rimanqakuchu,rimay,V+FR="parler"+FLX=V_SPP1+F+p+3
rimanqakus,rimay,V+FR="parler"+FLX=V_SPP1+F+p+3
rimanqakumá,rimay,V+FR="parler"+FLX=V_SPP1+F+p+3
```

3.2 IPS Mono Suffixation

On the other hand, the mono suffixation of verb stems by IPS suffixes are very productive compared to the PPS derivations. They give rise to new verbs as in these examples:

asiy	«laugh»	> asi-ri-y	to smile
ripuy	«go away»	> ripu-ku-y	to move
rakiy	«to split»	> raki-naya-y	to feel like splitting

The resulting forms can be conjugated as if it were a simple verb. Figure 3 shows the NooJ grammar that generates them.

uechua morphological grammar consists of 2 graphs.

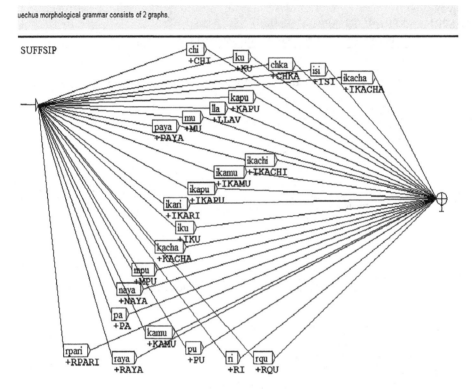

Fig. 3. The NooJ grammar that generates derived verbs using one suffix IPS

Below, we present an extract

rimaruy, rimay, V + FR = "parler" + FLX = V_SIP1_INF + PRES + INF
rimarquy, rimay, V + FR = "parler" + FLX = V_SIP1_INF + PAPT + INF
rimariy, rimay, V + FR = "parler" + FLX = V_SIP1_INF + DYN + INF
rimarayay, rimay, V + FR = "parler" + FLX = V_SIP1_INF + DUR + INF
rimapuy, rimay, V + FR = "parler" + FLX = V_SIP1_INF + APT + INF
rimapayay, rimay, V + FR = "parler" + FLX = V_SIP1_INF + FREQ + INF
rimanayay, rimay, V + FR = "parler" + FLX = V_SIP1_INF + ENV + INF
rimakuy, rimay, V + FR = "parler" + FLX = V_SIP1_INF + AUBE + INF
rimakapuy, rimay, V + FR = "parler" + FLX = V_SIP1_INF + RAS + INF
rimakachay, rimay, V + FR = "parler" + FLX = V_SIP1_INF + ARO + INF
rimaysiy, rimay, V + FR = "parler" + FLX = V_SIP1_INF + COLL + INF

The Quechua morphology allows certain agglutinations of two V_SIP2_INF and three V_SIP3_INF interposition suffixes. These new forms are also conjugable ones. They can be obtained by applying the following paradigms programmed in NooJ.

```
V_SIP2_INF  =<B>(:CHICHI  |:CHICHKA  |:CHIIKACHI  |:CHIIKAMU
|:CHIIKAPU |:CHIIKARI |:CHIIKU |:CHIISI  |:CHIKAMU |:CHIKU |:CHILLAV
|:IKACHAKAMU |:IKACHAKU |:IKACHALLAV |:IKACHAMU |:IKACHAPU
|:IKACHARIV     |:IKACHARQU     |:IKACHICHKA     |:IKACHIIKAMU
|:IKACHIIKAPU |:IKACHIIKARI |:IKACHIISI |:IKACHIKAMU |:IKACHIKAPU
|:IKACHIKU |:IKACHILLAV |:IKACHIMU |:  ...  |:RIVRQU  |:RPARICHI
|:RPARIIKACHI   |:RPARIIKAMU   |:RPARIIKAPU   |:RPARIIKU|:RPARIISI
|:RPARIKAMU |:RPARIKAPU |:RPARIKU |:RPARILLAV |:RPARIMU |:RPARIPU
|:RPARIRQU
   |:RQUCHKA  |:RQUISI  |:RQULLAV  |:RUCHKA  |:RULLAV  |:TAMUCHI
|:TAMUCHKA      |:TAMUIKACHA      |:TAMUIKACHI      |:TAMUIKAMU
|:TAMUIKAPU |:TAMURQU)y/INF;
```

It contains more than 240 valid agglutinations.

And for three IPS agglutinated suffixes

```
V_SIP3_INF =<B>(:CHAKUCHKAIKACHI |:CHAKUCHKAISI
|:CHAKUIKACHACHI |:CHAKUIKACHACHKA |:CHITAMURQU |:
|:CHKAIKACHIRIV |:CHKAIKACHITAMU |:CHKAISICHI |:CHKAISIIKARI
|:ISIMULLAV |:ISIMURQU |:KACHACHICHKA |:KACHACHIIKACHI
|:KACHACHIIKAMU |:KACHACHIIKAPU |:KACHACHIIKARI
|:KAMUIKAPUISI |:KAMUIKAPUKACHA |:KAMUIKAPULLAV |:KUIKUISI
|:KUIKULLAV |:KUIKURQU |:PAYAKAPUCHKA |:PAYAKAPUIKACHA
|:PAYAKAPUIKU |:PAYAKAPUISI |:......... .... |:RPARICHIMU |:RPARICHIPU
|:RPARICHITAMU |:RPARICHKAIKACHI |:RPARICHKAISI |:TAMUCHIKAMU
|:TAMUCHIKU |:TAMUCHILLAV |:TAMUCHIMU |:TAMUCHIPU
|:TAMUCHIRPARI |:TAMUCHKAIKACHI |:TAMURQUISI
|:TAMURQULLAV)y/INF;
```

Which contains more than 2470 valid agglutinations.

We present below some examples of ternary and more IPS suffix layer agglutinations:

na-ku-rqa rimanakurqanchik "we had talked about"
ra-ya-chi rimarayachinki "you have made him talk for a long time"
ra-ya-chi-spa rimarayachispayki "motivating him, you have made him talk for a long time"
ri-ku-lla-chka-pti rimarikullachkaptinchik "whereas we were talking very courteously one to each other"

3.3   The Mixed Inflections IPS + PPS

As we have seen in Fig. 1, an important property of Quechua grammar is that we can mix these two types of suffixation to generate current verbal forms. We have been able to program, with NooJ, the following paradigms of mixed agglutination:

> V_MIX1=(:SIP1_PR_V) (:SPP1_V) | (:SIP1_PR_C)(:SPP1_C)
> | (:SIP1_PRM_V) (:SPP1_V) | (:SIP1_PRM_C)(:SPP1_C);

  e.g.: miku-chi-**ni**-raq

> V_MIX12=:SIP1_PR_V)(:SPP2_V)|(:SIP1_PR_C)(:SPP2_C) |(:SIP1_PRM_V) (:SPP2_V)|(:SIP1_PRM_C)(:SPP2_C);

  e.g.: miku-chka-ni-raq-mi

> V_MIX21= (:SIP2_PR_V)(:SPP1_V)|(:SIP2_PR_C)(:SPP1_C);

> V_MIX22= (:SIP2_PR_V)(:SPP2_V)|(:SIP2_PR_C)(:SPP2_C);

e.g.: miku-chi-chka-ni-raq-mi

Where the symbols stand for V_MIX12: mixed verbal agglutination of 1 IPS and 2 SPP.

SIP1_PR_V: Agglutination with 1 IPS and conjugated following the PR scheme, etc.

We obtain, for the case of the verb *rimay*/to talk, 289 413 mixed and conjugated forms as shown in Fig. 4.

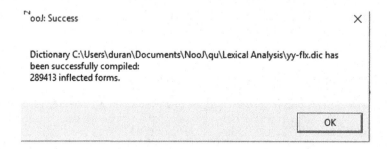

Fig. 4. Total of mixed and conjugated forms for the verb *rimay*

3.4   Verbalizers

An interesting strategy of Quechua is to generate verbs using a set of verbalizer suffixes. They may be added to nouns as well as to adjectives; these are:

The infinitive –*y*,
The transformative (creation or destruction) transitive -*chay*,

The transformative (to become) intransitive –*yay*.

Each one can be suffixed to certain classes of nouns or adjectives. Some nouns can be verbalized by only one of these suffixes and some nouns accept verbalization with two or all three of them: e.g.:

qipi « parcel »	→ qipi-y	«to carry a parcel »
pirqa « wall »	→ pirqa-y	« to build a wall »
machu « old man »	→ machu-yay	« to age »
tuta « night »	→ tuta-yay	« to get dark »
wasi « house »	→ wasichay	« to build a house »
wasi « maison, habitation »	→wasi-yay	« to become a house »
taksa « petit »	→ taksa-yay	« to get small »

The following NooJ grammars may be utilized to generate verbs:

YAY = *yay*/V; # derives some type of nouns and adjectives into verbs *chukru* =>*chukruyay* to toughen
CHAY = *chay*/V; # derives some type of nouns and adjectives into verbs *puka* => *pukayay*/to become red
Y = y/V;

3.5   The Quechua-French Translation Problem

One of the objectives of our long-term project, of machine translation from French to Quechua, is to build bilingual French-Quechua dictionaries of each PoS (Nouns, Verbs, Adjectives, etc.).

Concerning the Quechua lexicon of simple verbs, it contains around 1500 entries. It is not enough to get the translation of the 25 000 French verbal senses contained in Dubois & Dubois-Charlier dictionary[5]. For instance, the verb "tourner"/to turn, has 27 meanings in LVF (Dubois and Dubois-Charlier 2007; François et al. 2007); how to find its translation knowing that the only possible equivalent simple verb in Quechua is *muyuy*.

4   Generation of ALUc

But, as we have just seen above, Quechua has a very productive strategy: Derivation V > V, by suffixation. We have programmed the following paradigms in NooJ and generated 43 394 new Quechua verbal senses (Fig. 5. *Total of mixed and conjugated forms for the verb* rimay), ALUc, that can be conjugated as a simple verb.

---

[5] See Dubois & Dubois-Charlier «Les verbes français LVF» This dictionary contains 25 609 verbal senses. It was finished in 1992 and thanks to Denis Le Pesant, who is in charge of its diffusion, it is available since 2007 in a formalized format through MoDyCo.

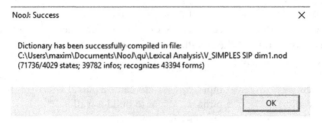

Fig. 5. Total of mixed and conjugated forms for the verb *rimay*

Total generated new verbal senses

An extract of which appear in the Fig. 6.

```
# NooJ V5
ichichiy,ichiy,V+TR+FR="marcher"+FLX=V_SIP1_INF+FACT+INF
ichirquy,ichiy,V+TR+FR="marcher"+FLX=V_SIP1_INF+PAPT+INF
ikiriy,ikiy,V+TR+FR="effleurer"+FLX=V_SIP1_INF+DYN+INF
iklliykuy,iklliy,V+TR+FR="bourgeonner"+FLX=V_SIP1_INF+COURT+INF
illariy,illay,V+TR+FR="absenter (s')"+FLX=V_SIP1_INF+DYN+INF
illapay,illay,V+TR+FR="absenter (s')"+FLX=V_SIP1_INF+PEAU+INF
illanayay,illay,V+TR+FR="absenter (s')"+FLX=V_SIP1_INF+ENV+INF
iñipay,iñiy,V+TR+FR="adhérer"+FLX=V_SIP1_INF+PEAU+INF
iñinayay,iñiy,V+TR+FR="adhérer"+FLX=V_SIP1_INF+ENV+INF
ipuykachay,ipuy,V+TR+FR="bruiner"+FLX=V_SIP1_INF+ARO+INF
ismuruy,ismuy,V+TR+FR="abîmer (s') (un fruit)"+FLX=V_SIP1_INF+PRES
kachamuy,kachay,V+TR+FR="envoyer"+FLX=V_SIP1_INF+ACENT+INF
kachiysiy,kachiy,V+TR+FR="saler"+FLX=V_SIP1_INF+COLL+INF
kachupay,kachuy,V+TR+FR="mâcher"+FLX=V_SIP1_INF+PEAU+INF
kakñakuy,kakñay,V+TR+FR="obstruer"+FLX=V_SIP1_INF+AUBE+INF
kallchariy,kallchay,V+TR+FR="moissonner"+FLX=V_SIP1_INF+DYN+INF
kamchachiy,kamchay,V+TR+FR="torréfier"+FLX=V_SIP1_INF+FACT+INF
kamiykuy,kamiy,V+TR+FR="admonester"+FLX=V_SIP1_INF+COURT+INF
```

Fig. 6. Extract of the generated new verbal senses

Do all of these ALUc have an actual meaning? In the affirmative case, the expected translation FR > QU of verbs of the Dubois & Dubois-Charlier dictionary seems to be feasible.

How to figure out the meaning of those derived forms which are not so familiar in everyday speaking or in the written corpus? When we apply the suffix –*chi*- to the verb *muyuy*/to turn we obtain *muyuchiy*/make it rotate, but is it possible to translate *muyu-chi-tamu-y* or *muyu-kapu-lla-y* avoiding long periphrases?

In order to aboard such a challenge, we have done the following steps: First we have identified and formalized the agglutination of IPS suffixes in the form of NooJ morphological grammars using one, two and three suffixes and applying some matrix methods to calculate the combinatorial. Second, we have inventoried the semantic characteristics of each IPS suffix and construct an indexed table. Third, we have constructed several NooJ grammars, like the ones appearing below, showing these modalities as annotations:

> CHAKU_T=chaku/RID_1_de   façon   rificule_RID_2_   de   façon   dimi-
> nuée_RID_3_de façon dépréciée;
>   CHI_T=chi/FACT_1_assiste_aide_FACT_2_invite_autorise_incite_FACT_3_pous
> se_l'objet_du_verbe_à_réaliser_l'action_FACT_4_commande_ordonne_pousse;
>   CHKA_T=chka/PROG1_en_train_deréaliser_l'action_PROG2_en_cours_d'accom
> plissement; # aspect progressif;
>   IKACHA_T=ykacha/DISP_1_en_désordre_DISP_2_réalise_l'action_desorienté_e
> n_directions_multiples_DISP_3_dans_n'importe_quel_manière;
>   ISI_T=ysi/COLL_1_collabore_COLL_2_aide_à_réaliser_l'action
>   RI_T=ri/DYN_1_commence_à_DYN_2_l'action_se_fait_une_fois_DYN_3_l'acti
> on_commence_lentement_et_à_la_legère_DYN_4_avec_courtoisie_et_amitié;

4.1 Modalities of Enunciation of IPS Suffixes

We have made an inventory of all the main modal and aspectual meanings of the 27 inter positional derivational suffixes SIP. We show in Fig. 7 an extract of these descriptions.

	SIP	formulle	VSV1	VSV2	VSV3
1					
2	chaku	chaku/RID	RID_1_de façon ridicule	RID2_de façon diminuée	RID_3_de façon dépréciée;
3	chi	chi/FACT	FACT_1_assiste_aide	FACT_2_invite_autorise_incite	FACT_3_pousse_l'objet_du_verbe_à_réal
4	chka	chka/PROG	PROG1_en_train_deréaliser_	PROG2_en_cours_d'accomplissement; # aspect progressif;	
5	ykacha	ikacha/DISP	DISP_1_le sujet réalise_acti	DISP_2_réalise_l'action_desorie	DISP_3_dans_n'importe_quel_manière;
6	ykachi	ikachi/POLI	POLI_1_poliment	POLI_2_concrètement	POLI_3_précisément
7	ykamu	ikamu/PREAT	PREAT_1_vers_le_sujet	PREAT_2_en_prévoyant	PREAT_3_attentionné
8	ykapu	ikapu/SOIN3	SOIN3_1_avec_attention	SOIN3_2_soigneusement	SOIN3_3_concernant_un_tiers
9	ykari	ikari/APRP	APRP_1_ponctuelle_et_rapi(	APRP_2_à_la_hâte_mais_avec_précision_réalise_l'action;	
10	yku	iku/COURT	COURT_1_courtoisement_so	COURT_2_allant_vers_l'intérieur	COURT_3_réalise_l'action_à_fond_se_lâc
11	ysi	isi/COLL	COLL_1_collabore	COLL_2_aide_à_réaliser_l'action;	
12	kapu	kapu/RAS	RAS_1_retour au sujet	RAS_2_réalise_l'action-en_auto_bénéfice	
13	ku	ku/AUBE	AUBE_1_actance_se_respons	AUBE_2_réalise_l'action_à_son_I	AUBE_3_s'impliquant_réalise_l'action
14	lla	lla/POL1	POL1_1_gentiment	POL1_2_poliment	POL1_3_doucement
15	mu	mu/ACT	ACT_1_actance_centripète_(	ACT_2_actance_centrifuge_d'él	ACT_3_action_orienté_vers_un_point;
16	mpu	mpu/INSP	INSP_1_l'action_se_réalise	INSP_2_rétrograder_de_retour_à_condition_antérieur;	
17	na	na/OBL	OBL-1_action_obligé	OBL_2_l'action_est_potentiellement_à_réaliser;	
18	naya	naya/ENV	envié_de_réaliser l action	ENV_2_bésoin_de_réaliser_l'action;	
19	pa	pa/PEAU	PEAU_1_peaufine_l'action	PEAU_2_réitère_l'action	PEAU_3_complète_l'action;
20	paya	paya/FREQ	FREQ_1_répétition_fréquent	FREQ_2_persiste_à_réaliser_l'ac	FREQ_3_action_prolongée_à_l'excès;
21	pu	pu/APT	APT_1_réalise_action_en_bé	APT_2_réalisé_en_préjudice_d'u	APT_3_action_s'écartant_du_sujet_et_de
22	ra	ra/PASS	RA_1: PRO V(PAS)	RA_2 : avoir_réalisée_l'action	
23	raya	raya/DUR	DUR_1_demeure_un_temps	DUR_2_permanence_dans_la_réalisation_de_l'action;	
24	ri	RIV=ri/DYN	DYN_1_commencer_à	DYN_2_l'action_se_fait_une_fois	DYN_3_l'action_commence_lentement_e
25	rpari	rpari/ASUR	ASUR_1_action_surprise_san	ASUR_2_réalise_l'action_impulsi	ASUR_3_action_réalisé_complètement;
26	rqu	rqu/PAPT	PAPT_1_accompli_l'action_e	PAPT_2_action_centrifuge_réalis	PAPT_3_brusquement_soudainement_de

Fig. 7. Modalities of enunciation of IPS suffixes

Parsing these grammars gives us several propositions as possible translations of each derived ALUc (Fig. 8):

```
rimapamuy,rimay,V+FR="parler"+FLX=V_SIP2_INF+le_sujet_peaufine_PEAU_2_réitère_PEAU_3_réalise_l_actio
rimapakuy,rimay,V+FR="parler"+FLX=V_SIP2_INF+le_sujet_peaufine_PEAU_2_réitère_PEAU_3_réalise_l_actio
rimapaisiy,rimay,V+FR="parler"+FLX=V_SIP2_INF+le_sujet_peaufine_PEAU_2_réitère_PEAU_3_réalise_l_acti
rimapaikuy,rimay,V+FR="parler"+FLX=V_SIP2_INF+le_sujet_peaufine_PEAU_2_réitère_PEAU_3_réalise_l_acti
rimapaikariy,rimay,V+FR="parler"+FLX=V_SIP2_INF+le_sujet_peaufine_PEAU_2_réitère_PEAU_3_réalise_l_ac
rimapaikapuy,rimay,V+FR="parler"+FLX=V_SIP2_INF+le_sujet_peaufine_PEAU_2_réitère_PEAU_3_réalise_l_ac
rimapaikamuy,rimay,V+FR="parler"+FLX=V_SIP2_INF+le_sujet_peaufine_PEAU_2_réitère_PEAU_3_réalise_l_ac
rimapaikachiy,rimay,V+FR="parler"+FLX=V_SIP2_INF+le_sujet_peaufine_PEAU_2_réitère_PEAU_3_réalise_l_a
```

Fig. 8. A glossed approach for the translation of derived verbs

For example: among the list of derivations of the verb *rimay*/to talk, it appears *rimaykuy*. It is the output of the inflection obtained when we apply to it the suffix –*yku*-. Its corresponding NooJ grammar has generated the following annotation in French:

```
rimaykuy,rimay,V+FR=« parler »+FLX=V_SIP1_INF_S+
le_sujet _courtoisement _COURT_2 soigneusement_
COURT_3_amicalement_ COURT_4_ réa-
lise_l_action_orienté_vers _le_sujet _+INF
```

We may use these automatic annotations in order to figure out the translation that feats better the ALUc.

We choose (manually): (the subject) talks to someone courteously, carefully, friendly,

Thus we'll have "to greet" as the nearest translation of *rimaykuy*.

Let's see an example in the sense French > Quechua. For one of the verbal senses of the verb "tourner"/to turn in French. We choose one of the 27 meanings of "tourner" appearing in the Dubois & Dubois-Charlier dictionary.

Meaning 13: tourner «Il indique que l'action se réalise de façon persistante»

To translate this verb, into Quechua, we search among the corresponding annotations in the output of inflections obtained when we apply the suffix –*paya*- to the verb *muyuy*, (that the modalities of *paya* appear in Fig. 7). Its corresponding NooJ grammar has generated, among 27 others, the following annotation in French:

```
muyupayay,muyuy,V+FR=« tourner »+FLX=V_SIP1_INF_S+
PAYA_1: PRO V(PR) sans répit, répétition fréquente _
PAYA_2: persiste_à_réaliser _l'action _ PAYA_3 : ac-
tion_prolongée_à_l'excés _ PAYA_F : « PRO V(F) sans répit
+INF
```

We choose manually: (the subject) "_tourne sans répit, répétition fréquente",

Which approaches better to meaning 13. Thus we'll have

muyupaayay: tourner de façon persistante (rotation)

We have mentioned before that the lexicon of simple verbs of Quechua contained around 1500 entries. We have found many derived verbs as translation of simple French verbs (like sourire/*asi-ri-y* coming from *asiy*/rire "to laugh"; bougonner/*rima-pa-ku-y* coming from *rimay*/parler "to talk", etc.). Many others come from verbalizations of nouns or adjectives like: aplanir/*pampa-cha-y* coming from the noun *pampa*/plane; appauvrir/*wakcha-ya-y*/*"to become poor"* coming from the adjectif *wakcha*/orphelin, pauvre "poor". However, for the translation of several thousands of French verbs we have utilized our annotation method described above. In Fig. 9 we present an extract of our newly constructed electronic French-Quechua verb dictionary containing more than 8600 entries (Duran 2017).

	A	B	C	D	E	F	G
1	V fr	V qu	Domaine	Phrase	CONJUGA	Construction	Lexique
2	aimer 01	kuyay	psycholo	On a~ P. Tous deux s'a~. On a~ sans être pay	1bZ	T1100 - P1	1
3	aimer 02	kuyaykuy	psycholo	On a~ ses amis pour leur franchise.	1bZ	T1100	5
4	aimer 03	kuyay	psycholo	On a~ manger des bonbons. On a~ le chocola	1bZ	T1400	5
5	aimer 04	waylluy	psycholo	On a~ cette rencontre, que tu réussisses, venir	1bZ	T1400	5
6	aimer 05	munay	psycholo	On a~ à taquiner sa soeur.	1bZ	N1a	5
7	aimer 06	munarpariy	physiolog	On a~ le soleil. Cette plante a~ l'ombre.	1bZ	T1306	5
8	ajouter 01	yapaykuy	littérature	On a~ un chapitre au livre.	1aZ	T13a0 - P3	1
9	ajouter 04	mirachiy	quantitati	Ces cris a~ à la panique.	1aZ	N3a	5
10	ajuster 01	matipay	mécaniq	On a~ un tuyau au robinet. La robe s'a~ bien à	1aZ	T13a8 - P3	2
11	ajuster 02	allchapariy	habilleme	On a~ sa cravate, sa coiffure.	1aZ	T1306	5
12	ajuster 04	tinkupachiy	mécaniq	L'ajusteur a~ les pièces de métal. La pièce es	1aZ	T1308 - P3	5
13	ajuster 05	tinkupay	habilleme	On a~ les salaires au coût de la vie.	1aZ	T13a0 - P3	5
14	ajuster 06	tinkuchiy	armemer	On a~ son fusil avant de tirer.	1aZ	T1306 - P3	5
15	ajouter	yapaykuy	mécaniq	On a~ un embout à une canalisation.	1aZ	T13a8 - P3	6
16	aligner 01	qayllascha	objet	On a~ des plants. Les chaises s'a~ devant l'es	1aZ	T1801 - P8	2
17	aligner 02	yupaychay	littérature	On a~ des nombres, des phrases vides.	1aZ	T1800	5
18	aligner 07 (s')	hayllascha	locatif, lie	Les gens s'a~ sur les trottoirs pour regarder le	1aZ	P7001	5
19	aligner 08 (s')	qawarikap	psycholo	On s'a~ sur les positions de P.	1aZ	P10g0	5
20	aligner 09	sqischay	locatif, lie	On a~ des enfants devant l'entrée de la classe	1aZ	T1701 - P7	5
21	alimenter 01	mikuy	cuisine, p	On a~ un malade de légumes. On s'a~ de fruits	1aZ	T11b0 - P1	2
22	alimenter 02	mikuchiy	industrie	On a~ en gaz la ville par une conduite. La ville	1aZ	T19j0 - P9(	5
23	alimenter 03	huntapay	littérature	On a~ les conversations d'anecdotes.	1aZ	T13b6 - P3	5
24	alimenter 04	huntapay	psycholo	On a~ sa haine de tels propos. Sa haine s'a~ (	1aZ	T1308 - P3	5
25	aliter 02 (s')	siririy	médecine	Le malade s'a~, on est a~ avec la grippe. La g	1aZ	P1001 - T9	5
26	aller 01	hamuy	locatif, lie	On a~ à Paris, dans un musée.	1zA	A12	1
27	aller 02	riy	locatif, lie	On a~ pour chasser. On a~ à la chasse, à la re	1zA	A12	5
28	aller 03	riykuy	commerc	On a~ chez le boucher, le médecin, le coiffeur.	1zA	A12	5
29	aller 04	riy	véhicule	Ce train a~ à Paris.	1zA	N3q	5

Fig. 9. Extract from Electronic French-Quechua (See Duran (2017).) verb dictionary

5   Conclusion and Perspectives

We have constructed several thousands of morpho syntactic paradigms for the inflection and derivation of verbs,

We have obtained the annotated verbal inflexions which serves as a database for the translation of several thousand french verbal senses into Quechua,

We plan to enhance the number of paradigms containing multi suffixed derivations to enlarge the FR-QU dictionary.

References

Dubois, J., Duboir-Charlier: Dictionnaire Linguistique et Sciences du langage. Editions Larousse, Paris (2007)

Duran, M.: Dictionnaire Quechua-Français-Quechua. Editions HC, Paris (2009)

Duran, M.: Formalizing quechua verbs inflexion. In: Proceedings of the NooJ 2013 International Conference, Saarbrücken (2013)

Duran, M.: Dictionnaire électronique français-quechua des verbes pour le TAL. Université Franche-Comté, Besançon-Paris. Thèse doctoral soutenu en mars, 600 p. (2017)

François, J., Le Pesant, D., Leeman, D.: Le classement syntaxico-sémantique des verbes français, Langue française, Armand Colin, Paris, no. 153 (2007)

Parker, G.J.: Ayacucho Quechua Grammar and Dictionary. University of Hawaii, Mouton, The Hague, Paris (1969)

Silberztein, M.: NooJ Manual (220 p., updated regularly) (2003). http://www.nooj4nlp.net

Silberztein, M.: La formalisation du dictionnaire LVF avec NooJ et ses applications pour l'analyse automatique de corpus. Langages 3/2010 (no. 179–180), pp. 221–241 (2010)

Silberztein, M.: Language Formalization: The NooJ's Approach. Wiley Eds. (2016)

Integrating the Lexicon-Grammar of Predicate Nouns with Support Verb *fazer* into Port4NooJ

Cristina Mota[1], Lucília Chacoto[2,3], and Anabela Barreiro[1(✉)]

[1] L2F/INESC-ID, Lisbon, Portugal
cmota@ist.utl.pt, anabela.barreiro@inesc-id.pt
[2] Universidade do Algarve, Faro, Portugal
lchacoto@ualg.pt
[3] CLUL, Lisbon, Portugal

Abstract. This paper describes the ongoing process of integrating approximately 3,000 predicate nouns into Port4NooJ, the Portuguese module for NooJ. The integration of these resources enables us to further extend the paraphrastic capabilities of eSPERTo paraphrasing system developed in the scope of a project with the same name. The integrated predicate nouns co-occur with the support verb *fazer* (*do* or *make*) and their syntactic and distributional properties are formalized in lexicon-grammar tables. These lexicon-grammar tables resulted in a standalone dictionary of predicate noun constructions and a few new grammars that can be used in paraphrase analysis and generation.

1 Introduction

Paraphrases are key linguistic assets in natural language interpretation and generation and play a very important role in NLP applications. Paraphrasing is a technique that implies the replacement or displacement of words, phrases or expressions in a sentence with semantically-equivalent linguistic structures so that the meaning of the output sentence remains the same. Support verb constructions represent an area of substantial interest in paraphrasing as they often have a corresponding verb associated with it. For example, the support verb construction *fazer uma sugestão* (*make a suggestion*), where the verb *fazer* has a weak semantic value and the predicate noun *sugestão* has a strong semantic value, is equivalent to the (strong) verb *sugerir* (*suggest*). Furthermore, support verb constructions are a rich source of paraphrases for reasons concerning stylistic variance allowed by the property of equivalence among different types of support verbs (elementary and non-elementary), among other reasons stated in early work (Barreiro 2009).

In this paper, we describe in detail the integration in the lexicon of Port4NooJ, the Portuguese module of NooJ (Silberztein 2016), of nearly 3,000 predicate nouns that co-occur with the support verb *fazer*, one of the most

S. Mbarki et al. (Eds.): NooJ 2017, CCIS 811, pp. 29–39, 2018.
https://doi.org/10.1007/978-3-319-73420-0_3

frequent verbs in Portuguese[1], thus complementing existing lexicon-grammar resources available for the Portuguese language. The resource integration work was developed within the eSPERTo[2] project whose main objective is to develop a smart, context-sensitive and linguistically enhanced paraphrasing system. At its current stage of development, eSPERTo comprises a paraphrase generator, a paraphrase acquisition module, and a web interactive application. eSPERTo recognizes semantico-syntactic, multiwords and other phrasal units, and transforms them into semantically equivalent phrases, expressions, or sentences. It uses local grammars to acquire linguistic knowledge that is applied in the identification/recognition and generation of different types of paraphrase, such as the support verb construction paraphrases described in Sect. 3. The new resources, which combine the dictionary of predicate nouns created by merging the information in lexicon-grammar tables with the entries in Port4NooJ plus the new transformational grammars contributed to over 5,000 entries before revision.[3] The utility of eSPERTo's paraphrasing capabilities has been explored in a question-answering system to increase the linguistic knowledge of an intelligent conversational virtual agent and in a summarization tool to assist the paraphrasing task, but the resources described in this paper have not been tested in these applications. The broader application envisaged for the paraphrasing system is e-learning, namely in helping Portuguese language learners in editing and revising their texts. Precise paraphrases can also be helpful in professional translation, editing, and proofreading, among other tasks.

2   Related Work

Support verbs have been extensively and systematically studied within the lexicon-grammar theory proposed by Gross (1975), from the theoretical and methodological points of view, for many languages. Practical analytic formalization of multiword units (some including support verb constructions) in the Romance languages have been presented by several authors (D'Agostino and Elia 1998; Laporte and Voyatzi 2008; Silberztein 1993), among others. Support verb constructions have also been taken into account in contrastive studies for English and French (Salkoff 1990). With regards to Portuguese, most studies

---

[1] Support verbs constructions with the verb *fazer* are used both in written texts and spoken language, occurring extensively in Portuguese sentences and noun phrases. Sentences with support verb constructions are more frequent than sentences with their equivalent verbs. In fact, Barreiro (2009) points out that from a search on all sentences of the COMPARA parallel corpus (Frankenberg-Garcia and Santos 2003; Santos and Inácio 2006) where the infinitive form of *fazer* occurs with a noun or with a left modifier and a noun, 47% of the times the occurrence is in a support verb construction.

[2] eSPERTo, which means 'smart' in Portuguese, stands for "System of Paraphrasing for Editing and Revision of Texts" ("Sistema de Parafraseamento para Edição e Revisão de Texto").

[3] Each entry in the lexicon-grammar can map to more than one existing Port4NooJ entry.

on support verb constructions outline the representation of argument mapping relations between a support verb and a nominalization (Baptista 2005; Chacoto 2005; Ranchhod 1990). Although support verbs often combine with autonomous predicate nouns, some studies focus on predicate adjective constructions (Carvalho 2007; Casteleiro 1981).

Mota et al. (2016) integrated the lexicon-grammar of human intransitive adjectives formalized by Carvalho (2007) into Port4NooJ, and showed how the properties contained in the lexicon-grammar tables can be used in paraphrasing tasks. The next step was to integrate complementary lexicon-grammars to expand the paraphrastic capabilities in Port4NooJ, building up on the previous work. Thus, the objective of the present work was to integrate the lexicon-grammar of predicate nouns which co-occur with the support verb *fazer* formalized by Chacoto (2005). The distributional and transformational properties of the predicate nouns addressed were used as the main criteria for establishing 19 classes and subclasses described in Sect. 3.

3   Lexicon-Grammar of Predicate Nouns with V_{sup} *fazer*

The systematic survey of predicate noun constructions with support verb *fazer* performed by Chacoto (2005) was carried out by assessing seven dictionaries and a subcorpus of five million words, i.e., Part 20 of the on-line 180 million word corpus CETEMPúblico.[4] The distributional and transformational properties of predicate nouns was used as the main criteria for establishing classes presented by these predicates, namely the syntactic structure, the number, and the type of constituents. Each class is represented in the form of a table, i.e., a binary matrix whose rows correspond to predicate nouns and whose columns represent the lexical and syntactic properties that these nouns have (+) or do not have (−). In total, 19 lexical and syntactic subclasses have been identified, as aforementioned.

The predicate nouns in the lexicon are, in general, everyday vocabulary (simple and compound nouns), with the exception of a group of predicate nouns of the sports and medical domains. The properties represented in the lexicon-grammar tables for each predicate noun lead to the generation of a variety of paraphrases, involving nominal group formation (active, passive and relative nominal groups), restructuring of the dative, shifting symmetrical nouns and symmetrical complements, conversion of the arguments, conversion into aspectual and stylistic variants of the support verb *fazer*, and transformation of the support verb, and relation between *fazer* and other support verbs. These properties establish semantic relationships between predicates allowing new features of paraphrases that have been described in NooJ grammars. For the sake of brevity, we illustrate only the case of paraphrasing based on the shifting of symmetrical nouns (Table 2), which are used with the predicates *aliar-se a* (*ally self with/to*), *casar com* (*marry with*), or *fazer um acordo com* (*make an agreement with*). The paraphrasing results from shifting the nouns from the position of subject to the position of indirect object and vice-versa.

---

[4] http://www.linguateca.pt/acesso/corpus.php?corpus=CETEMPUBLICO.

Table 1. Classes and distributional properties of predicate nouns with V_{sup} *fazer* (N_0 *fazer* Det (Npred + C) W)

Class	N_0 / Nhum	N_0 / E	Det / E	NPred / V-n, Adj-n	C / ≠V-n, ≠Adj-n	W / E	W / Prep N / Prep =:a	W / Prep N / Prep =:de	W / Prep N / Prep =:≠a, de	W / Prep Que-F	W / de N Prep N / Prep =:a	W / de N Prep N / Prep =:≠a, de	Example
FC					+	+							O Zé fez **voto de celibato**
FCAN					+		+						O Tó fez uma **serenata** à Ana
FCDN					+			+					O advogado fez as **alegações finais** do processo
FCPN					+				+				O avião fez uma **aterragem forçada** em Madrid
FCSI					+				+				O Zé fez um **acordo** com a Ana
FCQ					+					+			O Tó fez a **fineza** de convidar a Ana
FN				+		+							A instituição fez uma **angariação de fundos**
FNAN				+			+						O Zé fez um **aceno** ao Tó
FND	+		+	+									O Zé faz **ciclismo**
FNDN-hl				+				+					A Ana fez uma **classificação** dos dados
FNDNa				+				+					O rei fez a **abdicação** do trono
FNDNh				+				+					A Ana faz o **acolhimento** dos convidados
FNPN				+					+				O Zé fez um **acrescento** na prova tipográfica
FNSI				+					+				O Zé fez uma **aliança** com o Tó
FNQ				+						+			A Maria faz **tenção** de ter filhos
FNDNAN				+				+			+		O Zé fez uma **adaptação** do romance ao cinema
FNDNPN				+				+				+	O Tó fez o **câmbio** das pesetas em liras
FNDNPNSI				+				+				+	A Ana fez a **mistura** da farinha com o açúcar

Table 2. Paraphrasing based on symmetrical nouns

Symmetry	Paraphrases
$[N_0 \text{ V } N_1]$	*O Pedro aliou-se (com + a) a Maria*
	Pedro joined/allied himself with/to Maria
$[N_1 \text{ V } N_0]$	*A Maria aliou-se (com + a) o Pedro*
	Maria joined/allied himself with/to Pedro
$[N_0 \text{ Vcop Adj } N_1]$	*O Pedro é casado com a Maria*
	Pedro is married to Maria
$[N_1 \text{ Vcop Adj } N_0]$	*A Maria é casada com o Pedro*
	Maria is married to Pedro
$[N_0 \text{ Vsup N } N_1]$	*O Pedro fez um acordo com a Maria*
	Pedro made an agreement with Maria
$[N_1 \text{ Vsup N } N_0]$	*A Maria fez um acordo com o Pedro*
	Maria made an agreement with Pedro

4   Integration of Lexicon-Grammar Tables in Port4NooJ

The integration of the lexicon-grammar tables formalizing the properties of nominal predicates into Port4Nooj is a two-step process: first, one needs to create a dictionary from the lexicon-grammar and merge it with the Port4NooJ, and, then, it is necessary to create grammars that make use of the syntactic and distributional information described in the lexicon-grammar to identify and relate equivalent predicate constructions. The second step in our case aimed at expanding eSPERTo's paraphrasing capabilities. Before describing the integration process, we start by mentioning our major challenges to date.

Given our prior experience integrating the lexicon-grammar tables of the human intransitive adjectives, we expected the integration of the lexicon-grammar tables for the predicate nouns with the support verb *fazer* to be straightforward. However, that was not what happened and we were faced with challenges at different levels.

The first one worth mentioning is that 65% of the predicate nouns already existed in Port4NooJ as nominalizations. These nominalizations derived from the corresponding verb, but did not contain detailed syntactic and distributional information as the one formalized in the lexicon-grammar tables. Nonetheless, these entries include relevant information that cannot be discarded, such as semantico-syntactic (SAL) information and English transfers. Another challenge faced in the integration process was that, in contrast with the lexicon-grammar tables of the human intransitive adjectives, which had the morphological equivalent verb and adjective explicitly mentioned in the table, the lexicon-grammar tables in discussion simply indicate whether the predicate noun has morphological equivalent predicates (the corresponding entry has a plus (+) sign), but neither the equivalent verb nor the equivalent adjective are explicit in the table.

Instead, they are listed co-occurring with additional information in one cross-reference table as an appendix, i.e., not having all information concentrated at a single site makes integration a more complex task. Additionally, there are 800 multiword predicate nouns comprised of a noun and an adjective, such as *transcrição integral* (*full transcript*) whose morphological equivalent is a multiword construction [verb + adverb], such as *transcrever integralmente* (*fully transcribe*). These cases need further revision, as the equivalent constructions need to be treated in a slightly different way. Finally, the spelling of a few predicate nouns in the lexicon-grammar were updated to comply with the Portuguese Ortographic Agreement[5], but Port4NooJ does not yet comply with this Agreement.

4.1   From Lexicon-Grammar Tables to NooJ Dictionaries

We were compelled to adjust and create new scripts to overcome those challenges, even if the general process to create the equivalent standalone dictionary remained identical. For each entry in a lexicon-grammar table we had to (i) convert the corresponding lexicon-grammar properties into NooJ dictionary attributes; (ii) either create a new dictionary entry with those attributes or add those properties to an existing Port4NooJ dictionary entry or entries; and (iii) add the new entries or the old entries merged with the lexicon-grammar properties to the standalone dictionary. Each table was converted into a dictionary separately, but then all dictionaries were combined in a single standalone dictionary.

Representation of Lexicon-Grammar Table Properties. Given a lexicon-grammar entry, the new dictionary entry lemma is the predicate noun formalized in the lexicon-grammar, its POS tag is N, and its inflection code (the FLX attribute) is looked up in Port4NooJ or assigned automatically in cases where the word does not yet exist in the dictionary. The entry also receives the following attributes: +Npred, which indicates that it is a predicate noun, +Vsup=fazer which indicates that the support verb of the noun is *fazer*, and +Table= <name_of_ the_LG_table> whose value is the name of the lexicon-grammar table where its distributional and syntactic properties are described.

Similar to what was done previously in Mota et al. (2016), for each different column in a lexicon-grammar table, a property +<name_of_ prop> was created. If the noun row is marked with the value +, then that property was added to the noun entry, after removing or replacing special characters (e.g. Vsupter was generated from the LG property Vsup=:ter, and NONnhum was generated from N0=:N-hum), with the exception of the properties V and Adj, which indicate whether a noun has a morphologically-related verb and adjective, respectively. For these attributes, if the row has a + sign, instead of simply adding +V and +Adj attributes, a script looks up the noun in a file that lists all the morphologically-related verbs and adjectives to a noun and tries to obtain a derivation between the noun and verb or between the noun and

---

[5] http://www.portaldalinguaportuguesa.org/main.html.

the adjective. If the pair(s) noun/verb and noun/adjective exist, then a derivation paradigm is created and the following attributes are added to the entry:
+DRV=N2V<drv_code1>:<flx_code1> and +DRV=N2A<drv_code2>:<flx_code2>,
respectively. The drv code is determined and formalized automatically by finding the radical between the noun and the verb or adjective. For example, the
noun *espuma* (*foam*) is associated with the corresponding verb *espumar* (*turn
into foam*) and the corresponding adjective *espumoso* (*foamy*) through derivation rules (N2V2=r/V and A2V14=<B1>oso/A), which adds an −*r* to the noun to
create the verb, and replaces the nominal ending −*a* with the adjectival ending −*oso*, respectively. As it happens with the inflection code of the lemma, the
inflection of the derived word (flx_code1 or flx_code2) is determined by con-
sulting Port4NooJ (in this case, FLX=FALAR for the verb and FLX=ALTO for the
adjective).

After all the attributes of a lexicon-grammar entry have been converted to
NooJ format, we need to integrate the entry with the remaining Port4NooJ
dictionaries.

Integration with eSPERTo Dictionary Entries. As previously mentioned,
we had to use two dictionaries in order to properly integrate the lexicon-grammar
entries into Port4NooJ: current version, i.e., the current version prior to integrating the lexicon-grammar entries, and old version, i.e., the dictionary version
prior to the current version before removing the entries marked with +Npred that
derive from verbs. In this way, even if the nominal predicate does not exist in the
current version, it can still inherit the Port4NooJ properties as long as it makes
part of the older version. There are 608 nominal predicates that only exist in
the older version. If the nominal predicate exists in both dictionaries (there are
840 nominal predicates in this situation), and the linguistic information is the
same in both dictionaries, the entry in the new standalone dictionary inherits
that information, otherwise it only inherits the information that exists in the
older dictionary as it means that the information in the current dictionary does
not correspond to a nominal predicate - otherwise it would have been removed
as well.

Given those two versions of the Portuguese dictionary, a new entry on the
standalone dictionary of predicate nouns which co-occur with support verb *fazer*
is created in one of the following situations.

Scenario 1: if the predicate noun or a predicate noun compliant with the pre-
Ortographic Agreement[6] does not exist in Port4NooJ dictionary (neither current
nor old version) then add the new entry created following the process described
in the previous section as is. For example, the following entries did not exist in
Port4NooJ and were created directly from the lexicon-grammar table:

---

[6] In order to properly look up for the nominal entries in Port4NooJ, we applied a
simple rule that removes from words in Port4NooJ the first consonant in *cç*, *ct* and
pt before comparing it to the nominal predicate in the lexicon-grammar. However,
if there are derivation rules associated to the entry we still need to make sure that
the rules do not introduce pre-agreement spelling.

```
batota,N+FLX=CASA+Npred+Vsup=fazer+Table=FN+NONhum+DetE+DetUMModif+Npred
    +DRV=N2V2:FALAR+DRV=N2A6:ALTO+GNNdeNO
rodagem,N+FLX=ANO+Npred+Vsup=fazer+Table=FNAN+NONhum+DetO+Npred+Prepa
    +N1Nnhum+VsupestarPrep+Vaspiniciar+Vaspprosseguir+Vaspconcluir
    +DRV=N2V27:FALAR
```

Scenario 2: if noun (or the compliant noun) entry (i) exists in the current version but not in the old version or (ii) it is the same as in the old version, then merge the lexicon-grammar properties with the current entry, as in:

```
curva,N+FLX=ANO+Set=56+Subset=280+EN=curve+Npred+Vsup=fazer+Table=FN+NONhum
    +NONnhum+DetUMModif+DetO+Npred+Vsupdar+Vaspiniciar+Vaspprosseguir
    +Vaspconcluir+DRV=N2V2:FALAR
```

All information up to +Npred already existed whereas after +Npred was obtained from the lexicon-grammar table.

Scenario 3: if noun (or the compliant noun) (i) exists in old version only or (ii) it exists in both, but current and old entries differ, then merge the lexicon-grammar properties with the old entry. For example, the noun *cruzamento* exists both in the current and old versions, but the entries differ. So, the lexicon-grammar attributes are only added to the old version, as the following 4 entries illustrate - the first three entries existed in the current version and the last entry only existed in the old verb, consequently that is the information to which the lexicon-grammar attributes were added:

```
cruzamento,N+FLX=ANO+AB+mot+EN=crossover
cruzamento,N+FLX=ANO+CO+recp+EN=frog
cruzamento,N+FLX=ANO+PL+nagcom+EN=crossings
cruzamento,N+FLX=ANO+PresPart+Set=68+Subset=551+EN=intersecting
    +Npred+Vsup=fazer+Table=FNPN+NONhum+DetE+DetUMModif+DetO+Npred
    +Preppara+N1Nhum+DRV=N2V16:FALAR+GNNdeNOPrepN1
```

Additionally, remove previous Npred related properties, and then remove corresponding nominalizations from current version.

Table 3. Distribution of nominal predicates with support verb *fazer* by table attribute after integration into Port4NooJ

Class	# Lemmas	Class	# Lemmas	Class	# Lemmas
FNDN-hl	2145	FN	365	FNDNPNSI	120
FNDNa	1097	FNDNAN	313	FNSI	117
FNAN	560	FNDNPN	289	FND	92
FNPN	544	FCPN	246	FCSI	67
FNDNh	485	FCAN	180	FNQ	29
FC	406	FCDN	166	FCQ	27
Total: 6206					

Finally, create inflectional (FLX) and derivational (DRV) codes and corresponding rules as needed, and check for missing inflectional (FLX) and derivational (DRV) codes.

We started by integrating 18 of the 19 tables[7]. The first attempt to convert the tables into a standalone Port4NooJ dictionary resulted in 6,205 nominal entries, corresponding to 1,918 different noun lemmas. Most entries already existed in Port4NooJ (55%), which corresponds to about a 7% increase in nominal entries in Port4NooJ from 11,719 different noun lemmas that existed in the previous version of Port4NooJ (i.e., before removing entries corresponding to nominal predicates that are nominalizations). In terms of predicate nouns only, there was an increase of 25%. In addition, 332 new derivational paradigms were automatically created. Table 3 shows the distribution of nominal predicates in the lemma dictionary by table attribute sorted by the most frequent table assigned.

4.2   From Lexicon-Grammar Tables to NooJ Grammars

Although we have a preliminary version of the standalone dictionary that still needs to be reviewed, we initiated the process of using the information in the lexicon-grammar tables of these constructions to do paraphrasing. In particular, we started by addressing the following issues: (i) Port4NooJ grammars already paraphrase support verb constructions (for example, paraphrasing the verbal construction with the nominal construction and vice-versa, or alternation between the support verb and other stylistic or aspectual verbs) - grammars are being updated with attributes from the new tables and are also being extended to take into account other paraphrases involving this type of constructions; (ii) the

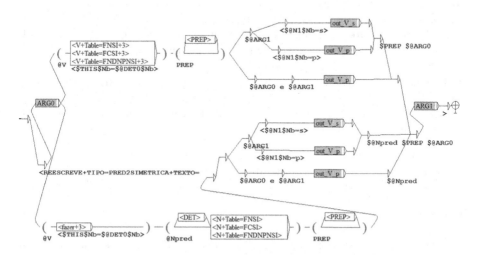

Fig. 1. Grammar to identify and paraphrase constructions that allow symmetry

---

[7] For now, the table that formalizes medical technical terms was left out for a second stage.

grammars that paraphrase active and passive constructions were only paraphrasing sentences involving a verbal predicate - we are modifying these grammars so that they can also paraphrase nominal predicates in sentences and noun groups.

In addition, we also started developing new grammars to paraphrase equivalent constructions based on specific properties formalized in those lexicon-grammar tables. One such grammar paraphrases symmetric predicates, such as those illustrated in Table 2 and pictured in Fig. 1. The grammar in Fig. 1 applied to the sentences *O homem apostou com a mulher* (*The man betted with the woman*) and *O homem fez uma aposta com a mulher* (*The man made a bet with the woman*) results in the paraphrase *O homem e a mulher fizeram uma aposta* (*The man and the woman made a bet*).

5   Conclusions and Future Work

This paper described the ongoing process and preliminary results of integrating the lexicon-grammar that formalizes the properties of predicate noun constructions with the support verb *fazer*. This process is only complete after consolidation of linguistic information in Port4NooJ dictionaries and grammars. Although we already had scripts to generate a standalone dictionary from the lexicon-grammar of adjectives, these scripts had to be modified to generate a new standalone dictionary from the lexicon-grammar of predicate noun constructions as the challenges faced during this integration were different from the ones encountered during the integration of the adjectives. We will continue this integration work by creating all necessary grammars to process the constructions formalized in the lexicon grammar tables in question. After completion of this integration task, we intend to revise and evaluate the new resources in distinct applications.

In the near future, we plan to continue integrating and adapting additional lexicon grammars, such as the constructions with the support verb *ser de* (*be of*), such as in *ser de uma ajuda inestimável* (*be of invaluable help*) formalized by Baptista 2005, as these are a rich resource for paraphrasing in Portuguese.

Acknowledgements. This research was supported by Fundação para a Ciência e Tecnologia (FCT), under exploratory project eSPERTo (Ref. EXPL/MHC-LIN/2260/2013). Anabela Barreiro was also funded by FCT through post-doctoral grant SFRH/BPD/91446 /2012. The authors would like to thank Max Silberztein for his prompt support and guidance with all matters related to NooJ.

References

Baptista, J.: Sintaxe dos predicados nominais com 'ser de'. Fundação Calouste Gulbenkian, Fundação para a Ciência e a Tecnologia, Lisboa (2005)
Barreiro, A.: Make it simple with paraphrases: automated paraphrasing for authoring aids and machine translation, Porto, Portugal (2009)

Carvalho, P.: Análise e Representação de Construções Adjectivais para Processamento Automático de Texto. Adjectivos Intransitivos Humanos. Ph.D. thesis, Universidade de Lisboa, Lisboa, Portugal (2007)

Casteleiro, J.M.: Sintaxe transformacional do adjetivo. INIC, Lisboa (1981)

Chacoto, L.: O Verbo Fazer em Construções Nominais Predicativas. Ph.D. thesis, Universidade do Algarve, Faro, Portugal (2005)

D'Agostino, E., Elia, A.: Il significato delle frasi: un continuum dalle frasi semplici alle forme polirematiche. In: AA. VV, Ai limiti del linguaggio, pp. 287–310. Laterza, Bari (1998)

Frankenberg-Garcia, A., Santos, D.: Introducing COMPARA: the Portuguese - English parallel corpus. In: Zanettin, F., Bernardini, S., Stewart, D. (eds.) Corpora in Translator Education, pp. 71–87. St. Jerome, Manchester (2003)

Gross, M.: Méthodes en syntaxe: régime des constructions complétives. Actualités scientifiques et industrielles. Hermann, Paris (1975)

Laporte, E., Voyatzi, S.: An electronic dictionary of French multiword adverbs. In: Language Resources and Evaluation Conference, Workshop Towards a Shared Task for Multiword Expressions, pp. 31–34 (2008)

Mota, C., Carvalho, P., Raposo, F., Barreiro, A.: Generating paraphrases of human intransitive adjective constructions with Port4NooJ. In: Okrut, T., Hetsevich, Y., Silberztein, M., Stanislavenka, H. (eds.) NooJ 2015. CCIS, vol. 607, pp. 107–122. Springer, Cham (2016). https://doi.org/10.1007/978-3-319-42471-2_10

Ranchhod, E.: Sintaxe dos predicados nominais com estar. Linguística, INIC, Lisboa (1990)

Salkoff, M.: Automatic translation of support verb constructions. In: Proceedings of 13th Conference on Computational Linguistics, COLING 1990, vol. 3, pp. 243–246. ACL, Helsinki (1990)

Santos, D., Inácio, S.: Annotating COMPARA, a grammar-aware parallel corpus. In: Calzolari, N., et al. (eds.) Proceedings of 5th LREC, Genoa, Italy, pp. 1216–1221 (2006)

Silberztein, M.: Les groupes nominaux productifs et les noms composés lexicalisés. Lingvisticæ Investigationes 17(2), 405–425 (1993)

Silberztein, M.: Formalizing Natural Languages: The NooJ Approach, p. 346. Wiley, Hoboken (2016)

Processing Agglutination
with a Morpho-Syntactic Graph in NooJ

Rafik Kassmi[✉], Mohammed Mourchid, Abdelaziz Mouloudi,
and Samir Mbarki

MISC, Ibn Tofail University, Kénitra, Morocco
rafik.kassmi@gmail.com, mourchidm@hotmail.com,
mouloudi_aziz@hotmail.com, mbarkisamir@hotmail.com

Abstract. In this paper, we have studied the morphological analyzer of Arabic in NooJ and we have contributed to its improvement in order to propose a new method of treatment of agglutination in Arabic. So, we will process the future tense like an agglutination and not like a conjugated tense using an inflectional graph. And we will cover all cases and possibilities of proclitic and enclitic respecting linguistic rules and we will transform the used morphological graph in NooJ to morpho-syntactic graph.

Keywords: Agglutination · Arabic language · NooJ · El-DicAr
Morpho-syntactic graph

1 Introduction

The morphological analysis of Arabic is based on the structure of the word. The richness of the structure of the Arabic word, its inflectional and derivational morphology and other morpho-syntactic phenomena such as non-voweling, agglutination and ambiguity, make its morphological analysis very complex [1, 2]. For decades, many morphological analyzers have been created but most of them remain incomplete, mainly due to the same reasons that we have already mentioned. In this study, we have decided to focus on the morphological analyzer of Arabic in NooJ (El-DicAr). Our goal is to improve this work by covering all cases of agglutination respecting linguistic rules, transforming the used morphological graph to morpho-syntactic graph and processing the future like agglutination.

2 A Brief Description of Arabic Language

Arabic is a Semitic language that is morphologically complex. "The general rule for Arabic is that everything is at least five times more complicated than any European language" [3].

The Arabic language consists of 28 letters of which 1 long vowel (ا‍ - [à]) and 2 semi-consonants (ي - [ya]) and (و - [wa]) named glides. According to their appearance in the sentence, the glides can be considered as consonants (وَعَد [wa'da]) or long vowels (دُرُوس [durûsun]).

© Springer International Publishing AG 2018
S. Mbarki et al. (Eds.): NooJ 2017, CCIS 811, pp. 40–51, 2018.
https://doi.org/10.1007/978-3-319-73420-0_4

Arabic also contains brief vowels called diacritics that are written above or below the consonants they follow َـ-a [فَتْحَة -fatha], ِـ-i [كَسْرَة -kasra], ُـ-u [ضَمَّة -damma], ْـ [سُكُون -sukûn], ّـ [شَدَّة -šadda]. Generally, the brief vowels do not appear in Arabic text. The concept *tanwīn* represents the indefinite words without article or complement, built by doubling the brief vowels (ًـ [an], ٍـ [in], ٌـ [un]) and placed at the end of the word.

The Arabic language is written from right to left. It is semi-cursive that all letters can be attached together and change shape depending on whether they are at the beginning, middle or end of the word. However, there are six letters that never attach to the letters that follow them (أ [à] – و[wa] – ر [ra] – ز [za] – د [da] – ذ [d̲]).

Arabic is a generative language where the majority of words (also called Lemmas) are derived from a root while respecting a scheme. This concerns verbs, nouns and some particles.

The word is a series of graphemes between two blanks that correspond to a form or a unit likely to appear under a lexical entry or lemma [4].

Arabic is a generative language where the majority of words are derived from a root (called also radical) while respecting a pattern. This concerns verbs, nouns and some particles. A root is formed by a succession of two, three or four letters but around 64% of the Arabic roots are made up of three letters [5]. In limited cases, especially for nouns, a root can have more than four letters. All these letters form the basis of the word [6]. A family of words can be generated from one root using different schemes [7] (Table 1).

Table 1. Example of generated words from the root [ktb]

Scheme	كتب[ktb]	Transliteration
فَعَلَ[fa'ala]	كَتَبَ[kataba]	He wrote
فَاعَلَ[faā'ala]	كَاتَبَ[kaātaba]	He wrote someone
فَاعِلٌ[faā'ilun]	كَاتِبٌ[kaātibun]	Writer
مَفْعَلٌ[maf'alun]	مَكْتَبٌ[maktabun]	Desk
فِعَالٌ[fi'aālun]	كِتَابٌ[kitaābun]	Book
مَفْعُولٌ[maf'ûlun]	مَكْتُوبٌ[maktûbun]	A writing
مَفْعَلَةٌ[maf'alatun]	مَكْتَبَةٌ[maktabatun]	Library
فِعَالَةٌ[fi'aālatun]	كِتَابَةٌ[kitaābatun]	Writing

There are three categories of Arabic words: verbs, nouns and particles. The verb is a word that in syntax conveys an action, an occurrence, or a state of being. The Arabic verbs are classified according to the number or nature of root consonants and also according to their schemes [8]. Derived verbs are conjugated with the same prefixes or suffixes like nude verbs. A trilateral verb cannot be increased by more than three letters, and a quadrilateral verb cannot be increased by more than two letters. Therefore, an Arabic verb cannot exceed six letters.

The noun refers to a being or an object that has an independent sense of time. There are conjugated nouns that are expressed in the singular, dual and plural. There are either primitive non-derivational nouns like قِرْدٌ [qirdun] (monkey), or derivational nouns from the root like كِتَابٌ [kitaābun] (book). There are also non-conjugated nouns that keep their forms regardless of the context.

The particle serves to locate events regarding time and space. They are generally considered as important particles that contribute to the coherence of the phrase without having a specific meaning to a specific area. They can have prefixes and suffixes that make their identification more complex [9].

3 Agglutination in Arabic Language

The Agglutination refers to complex words formed by a set of morphemes bonded to each other and giving several morpho-syntactic information. Arabic is a strongly agglutinative language that is why its automatic processing is very complex and difficult.

In Arabic, each lexical unit can be decomposed as proclitic, prefix, lemma, suffix and enclitic. For example, the word أَوَسَتأكلونه (awasataàkulûnahu), which means *will you eat it?* is decomposed like (Table 2):

Table 2. Example of segmentation

هُ	ونَ	أكُلُ	تَ	أوَسَ
hu	ûna	àkulu	Ta	à+wa+sa
It		Eat		Will you
Enclitic	Suffix	Lemma	Prefix	Proclitic

Arabic linguists have created a set of rules, concerning order and compatibility between morphemes, which must be respected in processing agglutination if the lexical unit is a verb, a noun or a particle.

3.1 Proclitics

The proclitics are prepositions or conjunctions (question, future, etc.). Table 3 presents all possible morphemes for proclitics before verbs. We can combine, at least three morphemes respecting order. Morphemes of the same position cannot be combined.

Table 3. Possible morphemes before verbs

1ˢᵗ position	2ⁿᵈ position	3ʳᵈ position		
Interrogation	Coordination	(near) Future	Subjunctive	Corroboration
à	wa - fa	sa	li	la
أ	وَ ـفَ	سَـ	لِ	لَ

Here are all possible morphemes for proclitics before nouns. We can combine 3 morphemes and, rarely, four morphemes respecting order (Table 4).

Table 4. Possible morpheme before nouns

1ˢᵗ position	2ⁿᵈ position	3ʳᵈ position	4ᵗʰ position
Interrogation	Coordination	Preposition	Definite article
à	wa - fa	ka - li – bi	al
أ	وَ ـفَ	بِـلِ ـكَ	ال

And here are all possible morphemes for proclitics before particles. We can combine two morphemes for particles (Table 5).

Table 5. Possible morpheme before particles

1ˢᵗ position	2ⁿᵈ position
Interrogation	Coordination
à	wa - fa
أ	وَ ـفَ

3.2 Prefixes and Suffixes

The prefix and suffix express grammatical features and indicate:

– Case of the noun. For example the word مُنْتِجَتَانِ (muntiğataāni/two women producers), the suffix تَّانِ (ataāni) indicate the subject case of noun مُنتِجٌ (muntiğun/producer) in duel-female.
– Verb Mode (active, passive)
– Numbers (singular, duel, plural)
– Gender (male, female)
– Person (1st, 2nd, 3rd).

For instance, in the word يَأْكُلُوا (yaàkulû/they eat) the prefix (يَ/ya) and the suffix (وا/û) indicate the conjugation of the verb أَكَلَ (àakala/to eat) in present active voice in third person at the plural-male (Table 6).

Table 6. Example of prefix and suffix of verb أَكَلَ (àakala/to eat)

3rd person	Male	Female
Singular	(يَ) أَكُلُ	(تَ) تَأْكُلُ
	yaàkulu / eats	taàkulu / eats
Duel	(يَ) يَأْكُلَا (ا)	(تَ) تَأْكُلَا (ا)
	yaàkulaā / eat	taàkulaā / eat
Plural	(يَ) يَأْكُلُوا (وا)	(يَ) يَأْكُلْنَ (نَ)
	yaàkulû / eat	yaàkulna / eat

3.3 Enclitics

The enclitic is a personal pronoun and it depends on person, gender and numbers. For example, the enclitic هَا (haā) of the word كِتَابُهَا (kitaābuhaā/her book) indicate the third person female singular of the word كِتَابٌ (kitaābun/book).

4   The Morphological Analyzer of Arabic in NooJ

NooJ is a platform of linguistic development and natural languages formalization used essentially in Automatic Natural language processing (ANLP) [10]. It is based on an integrated architecture gathering a set of linguistic modules namely spelling, corpus analysis and annotation, flexion, morphology, derivation, syntax, semantics and ontology.

NooJ also integrates a graph editor, a concordancer, a grammar debugger and statistical tools. NooJ allows its users to build, test and manage dictionaries and grammars in text or graph forms. It is equipped with a robust linguistic engine that allows analyzing and processing the specificities of twenty languages such as Arabic, Chinese, Dutch, French and other languages. All these resources are available for free on the NooJ official web site.

El-DicAr is an automatic morpho-syntactic analyzer for Arabic language created in NooJ to recognize named entities. It is a Lemma-Based dictionary. It proposes a tokenization system of agglutinated forms based on two phases. Firstly, it applies a decomposition of all agglutinated forms with morphological grammars. Then, it applies morpho-syntactic rules to all morphemes [1] (see Fig. 1).

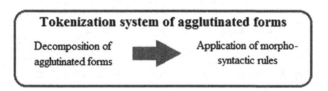

Fig. 1. Tokenization system in NooJ

5   Contribution

In this paper, we will improve the morphological analyzer of Arabic in NooJ (El-DicAr) by developing four points.

5.1   Processing the Future

In El-DicAr, the future is processed like a tense by adding the morpheme (سـ/sa) as prefix to the present tense form (active and passive voice) generated by an inflectional paradigm (see Fig. 2).

For example in Fig. 3, the word سَيَأْكُلُ (sayaàkulu/he will eat) is annotated in NooJ as أَكَلَ, V + Synt = Tr + Temps = F + Pers = 3 + Genre = m + Nombre = s.

This will generate additional entries in the dictionary (about 136,500 entries for 10,500 verbs). Which generates a voluminous dictionary and of course more time to load all resources.

Our proposition is to process the future with agglutination by using the morpheme (سـ/sa) as a prefix that joins verbs in present active and passive voice (see Fig. 4).

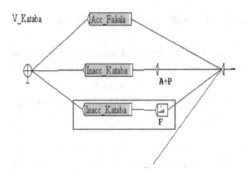

Fig. 2. Example of inflectional graph used in El-DicAr

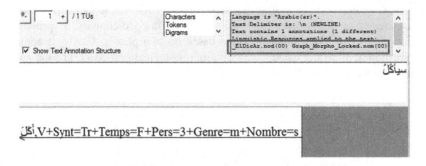

Fig. 3. Example of recognition of the future with El-DicAr

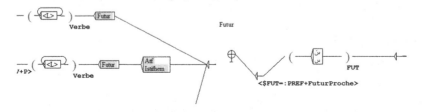

Fig. 4. Morphological graph processing the future as agglutination

For example in Fig. 5, the word سَيَأْكُلُ (sayaàkulu/he will eat) will be annotated in NooJ as morpheme سـ, PREF + FutureProche followed by the verb يَأْكُلُ (yaàkulu/he eats) in present (أَكَلَ, V + Synt = Tr + Temps = P + Pers = 3 + Genre = m + Nombre = s).

5.2 Rules

Agglutination Rules. In El-DicAr, the used morphological grammar does not deal with some cases even though it respects the linguistic rules. In the example (see Fig. 6), no annotations are given to recognize the word أَفَلِيَأْكُلُ (àfaliyaàkula/is it for him to eat?)

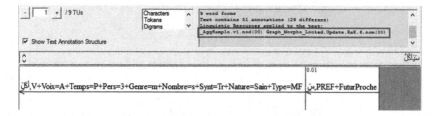

Fig. 5. Example of recognition of the future with agglutination

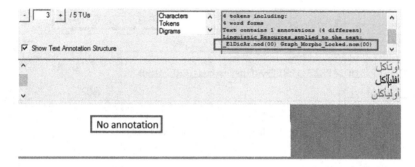

Fig. 6. Example of non-recognized word with El-DicAr

So, we tried to cover all possibilities by combining all possible morphemes respecting order and position. And with our morphological grammar, the word أَفَلِيَأْكُلَ (àfaliyaàkula/*is it for him to eat?*) will be recognized as follow (see Fig. 7).

Fig. 7. Recognition of the word أَفَلِيَأْكُلَ with the new morphological graph

And the applied rule is "Interrogative *particle* + *Coordination particle* + *Corroboration particle* + *Verb*".

Here are some examples of additional rules that we cover in our graph

- Interrogative particle + Coordination particle + Future particle + Verb + personal pronoun. Word: أَوَسَتَأْكُلُونَهُ (àwasataàkulûnahu/*Will you eat it?*) (see Fig. 8)
- Interrogative particle + Coordination particle + Particle. Word: أَوَفِيهِمْ (àwafihim/*is there any among them?*) (see Fig. 9).

Fig. 8. Recognition of the word أَوَسَتَأْكُلُونَهُ

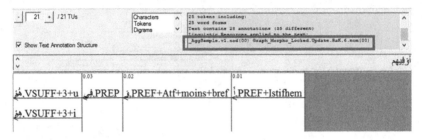

Fig. 9. Recognition of the word أَوَفِيهِمْ

Transitivity Rules. In Arabic, there are two categories of verbs (see Fig. 10): transitive (مُتَعَدِي/muta'addi) and intransitive (لَازِم/laāzem, for example: هَاجَ البَحْرُ/haāǧa albahru/the sea surged). Transitive verbs can have one direct object (for example: أَكَلَ عَلِيٌالتُّفَاحَة/akala aliun attuffaāhata/Ali ate the apple), or two direct objects (example: حَسِبَ عَلِي اَلخُبزَ جَاهِزا/hasiba Aliun alkhubza ǧaāhizan/Ali deemed the bread ready) and rarely three direct objects.

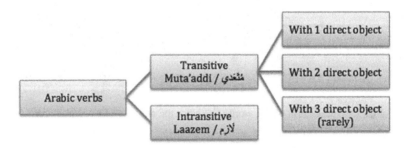

Fig. 10. Transitivity in Arabic verbs

In El-DicAr, the used morphological graph does not recognize transitive verbs with two objects. So, in our work we have succeeded to treat all cases of transitive verbs.

For example, let us consider the longest word in the holy Quran [فَأَسْقَيْنَاكُمُوهُ/faàsqaynaākumûhu/and given you drink from it] (*Surah N 15 «Al-Hijr» verse 22*). Here are the annotation result of this word (see Fig. 11).

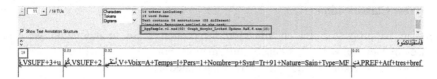

Fig. 11. Recognition of the word فَأَسْقَيْنَاكُمُوهُ

In our work, we transform the used morphological graph to morpho-syntactic graph by enriching annotations and proposing a new classification.

5.3 Annotations

Our goal is to give the maximum of details to all recognized morphemes even some syntactic information. So, annotations become rich and more explicit. For example, the word [وَسَيَأْكُل/wasayaàkulu/and he will eat] is recognized with El-DicAr, which only focuses on "important" categories, like the annotation of the morpheme [و/wa/and] is CONJ (conjunction) (see Fig. 12).

Fig. 12. Example of annotation with El-DicAr

With our graph we try to give more syntactic information to morphemes. In this example, the morpheme [و/wa/and] is annotated as [PREF + Atf + moins + bref] which is a shorter conjunction prefix (see Fig. 13). In this case we have an additional morpheme [س, PREF + FutureProche/prefix for near future] because we process the future tense with agglutination.

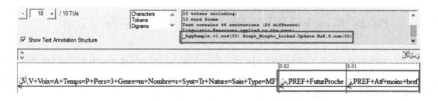

Fig. 13. Example of annotation with the new morphological graph

5.4 New Classification

Moreover, we propose a new classification of morphemes. So, in annotation, we give all cases of morphemes with syntactic information. For example in Fig. 14, the sentence [لَايَأْكُل/laā yaàkulu/He does not eat] is annotated as

[أَكَلَ,V+Synt=Tr+Voix=A+Temps=P+Pers=3+Genre=m+Nombre=s).] +

[لَا, ADV+HARF]

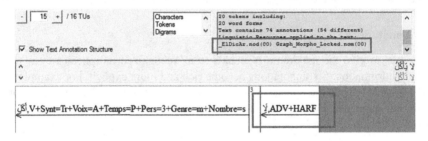

Fig. 14. Example of classification with El-DicAr

With our morpho-syntactic graph, we give the two possible syntactic cases of morpheme: the imperative negation [لَا, ADV + Naahiya] and the absolute negation [لَا, ADV + Naafiya]. (See Fig. 15).

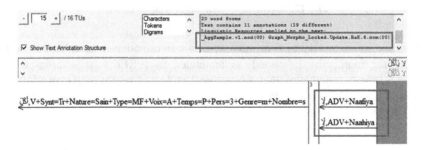

Fig. 15. Example of classification with the new morphological graph

Finally, as illustrated, we transform the used morphological graph in El-DicAr into a morpho-syntactic graph. So simultaneously, we decompose the lexical unit and give the corresponding syntactic annotation for each morpheme.

6   Conclusion and Perspectives

In this paper we have described the agglutination in the Arabic language and briefly presented the linguistic rules of agglutination. Then, we have contributed to the improvement of the morphological grammar of El-DicAr. Firstly, we have proposed the processing of the future tense with agglutination. Then, we have tried to cover all cases and possibilities of proclitics and enclitics respecting linguistic rules. Then, we have transformed the used morphological graph into a morpho-syntactic graph by enriching annotations and proposing a new classification.

Our prospective work will further improve this morphological grammar to process other categories of prefixes, and develop the structural grammars.

References

1. Mesfar, S.: Analyse morpho-syntaxique automatique et reconnaissance des entités nommées en arabe standard. Ph.D. thesis, Franche-Comte University, France (2008)
2. Boulknadel, S.: Traitement automatique des langues et recherche d'information en langue arabe: Apport des connaissances morphologiques et syntaxiques pour l'indexation. Ph.D. thesis, Nantes University, France (2008)
3. Beesley, K.: Finite-state morphological analysis and generation of Arabic at Xerox research: status and plans in 2001. In: Actes du Workshop the Arabic Language Processing. ACL Workshop, Toulouse (2001)
4. Tuerlinckx, L.: La lemmatisation de l'arabe non classique. In: 7ème journées internationales d'analyse statistique des données textuelles, JADT 2004, Louvain-la-neuve, Belgique (2004)
5. Khoja, S., Garside, R., Knowles, G.: A tagset for the morphosyntactic tagging of Arabic. In: Proceeding of Corpus Linguistics, pp. 341–353, Lancaster, UK (2001)
6. Abdelwahed, "العربيالتّصريففيقراءةالفعلبنية". Sfax University, Tunisia (1996)
7. Douzidia, F.S.: Résumé automatique de texte arabe. Master thesis, Montréal University, Québec (2004)
8. https://en.wikipedia.org/wiki/Arabic_verbs
9. Khemakhem: ArabicLDB: une base lexicale normalisée pour la langue arabe. Master thesis, Sfax University, Tunisia (2006)
10. Silberztein, M. (ed.): Formalizing Natural Languages: The NooJ Approach. Wiley-ISTE, January 2016. ISBN 978-1-84821-902-1

Formalizing Arabic Inflectional and Derivational Verbs Based on Root and Pattern Approach Using NooJ Platform

Ilham Blanchete[(⊠)], Mohammed Mourchid, Samir Mbarki,
and Abdelaziz Mouloudi

MIC Research Team, Laboratory MISC, IbnTofail University, Kenitra, Morocco
ilham.blanchete@gmail.com, mourchidm@hotmail.com,
mbarkisamir@hotmail.com, mouloudi_aziz@hotmail.com

Abstract. This article presents the inflectional and the derivational model of Arabic verbs based on root and pattern approach, using a linguistic classification that determines and specifies a set of morphological properties.

Our work in NOOJ platform is based on:

- A dictionary that consists of roots, lemmas and patterns.
- Generating all possible verbs inflectional and derivational forms using our NooJ grammars.

Since formalizing Arabic verb's requires the determination of certain unavoidable morphological properties as root and pattern, we started by specifying them; the matching process between them gives twenty thousand verbs entries [1]. Consequently, this work stands on the following principles:

- Categorizing all possible roots and patterns.
- Matching roots with patterns that give Lemma, which is a verb in our case.

Our dictionary considers, in addition to roots and patterns, lemmas as dictionary entries, which allows making an advanced search in a text.

Root and pattern serve to build the meaning of most Arabic word [1]. For instance, matching the root(KTB-كتب) with the pattern فَعَلَ [Fa3aLa], gives the lemma (verb): كَتَبَ(KaTaBa-to write); matching the same root with the pattern فِعَال(Fi3aLun) gives the lemma (noun): كِتَابٌ(KiTaBun – a book); matching the same root with different patterns gives different Arabic words.

The implementation of this work is a dictionary that contains all inflectional and derivational verbs forms.

Adding other Arabic words (nouns, adjectives, infinitives) with their inflectional and derivational forms, in order to complete our dictionary, is considered as our future perspective.

Keywords: ANLP · NooJ platform · Inflection and derivation

© Springer International Publishing AG 2018
S. Mbarki et al. (Eds.): NooJ 2017, CCIS 811, pp. 52–65, 2018.
https://doi.org/10.1007/978-3-319-73420-0_5

1 Introduction

The linguistic analysis must go through a first step of lexical analysis, which consists of testing the membership of each word of the text to the Arabic vocabulary. So, we must begin with the formalization of the Arabic vocabulary [2]. Verbs cover a large part of Arabic vocabulary. Each Arabic verb has a one single root, also called radical, and a single pattern. In addition to these two components, the verb has other morphological properties that we are going to detail.

The verb is classified into several categories according to the number and the nature of its root letters. The verbs that have 3 letters (called trilateral verbs) form 64% of Arabic verbs, like the verb (to write - كَتَبَ- KaTaBa), those that have 4 letters (called quadrilateral verbs) form 33% of Arabic verbs, like the verb (to assure - طَمْأَنَ - TaMAaNa), and those that contain 5 letters, form the rest of the percentage [3], the most frequently ones are those that have 3 radical letters.

In this paper we give an overview about the first and the only dictionary that was implemented in NooJ platform (EL-DiCar), then we present a theoretical study that reflects the linguistic side of our work, and answers the question: why do we have to specify the root and the pattern during the formalization of Arabic vocabulary especially verbs? And, We also give the linguistic classification of Arabic verbs that we have adopted during the formalization process, as a first part, the second part is the practical study which is the implementation of the theoretical study that was enumerated in the first part of the article, using NooJ platform importantly; we are going to detail the dictionary structure, and give an example of our own lexical grammar that we have implemented to generate the inflectional and the derivational forms of all dictionary entries. In addition to that we are going to present the annotations as the result of the text linguistic analysis. Finally, we are going to present the comparison results between the search operation using root and pattern in our dictionary and the lemma search in the previous dictionary.

2 Related Works

As far as the previous works in NooJ platform are concerned, there is only one dictionary called the Arabic EL-DiCar dictionary that is based on lemma approach. EL-DiCar retrieves all inflectional and derivational forms of one entry using lemma search. It contains, in addition to the Arabic words, 10375 fully vowelled entries of verbs; each entry represents third person, singular, masculine and perfect verbs [4]. These entries are not related to each other even if they share the same root. For example: the verb (to write – KaTaBa – كَتَبَ), the verb (to write a letter - iKTataBa – اِكْتَتَبَ), the noun (library – maKTaBa - مَكْتَبَة) and the noun (book – KiTaB- كِتَاب) share the same root that is: [ktb-كتب], but they are not linked to each other in this dictionary; entries have no relation between them whereas the dictionary is based on lemma. This limits the search operation, and makes the application of several operations on the text like retrieving concepts or the auto-correction more complex. Our model that is based on root and pattern approach retrieves all entries with their inflectional and derivational forms that share the same root. It can also we can extract concepts within a text using pattern search even if these concepts do not share the same root.

3   Theoretical Study

3.1   Arabic Verb Morphology

Arabic morphology exhibits rigorous and elegant logic. It consists primarily of a system of consonant roots, which interlocks with pattern to form most Arabic words, especially verbs [5]. With the same root we can derive several words by using different patterns. In this context it is necessary to provide a definition of root and pattern:

A root is a relatively invariable discontinuous bound morpheme, represented by two to five phonemes, typically three consonants in a certain order, which interlocks with a pattern to form a stem [5]. For instance, the verb (to write, KaTaBa - كَتَبَ) has the root letters (KTB-كتب), that are interlocked with the pattern (FaAaLa – فَعَل). To clarify, the first letter in the root will be substituted with the first letter in the pattern and so on. This process of interlocking aims to form the verb (to write, KaTaBa-كَتَبَ), as well as the quadrilateral roots like the root (TMAN- طمأن) and the pattern (FaALaLa-فَعْلَل) that gives the verb (to assure –TaMAaNa - طَمْأَنَ).

The importance of specifying the roots and the patterns in the morphological analysis during building a dictionary, and adopting them as dictionary entries is to make all nouns, verbs and adjectives; in general all Arabic words that share the same root or concepts, retrievable by their root or pattern.

A pattern is a bound and in many cases, discontinuous morpheme that carries meaning [6]. For example, the pattern (FaAaLa-فَعَل) refers to an action, matching this pattern with:

- The root (KTB-كتب) gives the verb (to write, KaTaBa –كَتَبَ).
- The root (SANAa-صنع) gives the verb (to make-SaNaAea-صَنَعَ).
- The root (JLS-جلس) gives the verb (to sit-JaLaSa-جَلَسَ).

With a simple search operation using this pattern, we can extract all verbs that refer to an action, even if they do not share the same root.

Matching the pattern (MaFALa-مَفْعَلُة) that refers to a place with:

- The root (KTB-كتب) gives the noun (a liberary - maKTaBa - مَكْتَبَة).
- The root (KHBZ-خبز) gives the noun (a bakery shop - miKHBaZa -مَخْبَزَةُ).

To extract all nouns that refer to a place, we simply apply a pattern search using the previous pattern, the same thing applies to all other patterns.

The following table represents a part of the Arabic lexicon that is formed by the possible matching process between roots and patterns: 1 means that the matching process is possible, while 0 means that the matching process is not possible or there is no Arabic lemma formed from this combination (Table 1).

3.2   Verb Classification

This linguistic classification covers all Arabic verbs categories [6]. Structured as a tree, each leaf represents a verb model that we are going to formalize using NooJ platform; Fig. 1 shows a part of the classification tree that we used.

Table 1. Capture of Arabic lexicon formed from matching several roots and patterns

Root/pattern	فَعَلَ FaAaLa	فَعُلَ FaAoLa	فَعِلَ FaAiLa	فَاعِلَ FaAiL	تَفَاعَلَ taFaALa
حصن HSN	0	1	1	1	0
كتب KTB	1	0	0	1	1
ضرب DRB	0	0	1	1	1

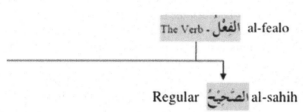

Fig. 1. Arabic regular verb classification

Arabic verbs are divided into two main classes: regular verbs, like the verb (to write), and weak verbs like, the verb (to say) [6]. We are going to detail one of them.

Regular Verbs. Verbs that their roots are free of vowel are also called (الأَفْعَالالأَصَحِيحَة.- AaFAaL - SaHiHa) like the verb (to write كَتَبَ – KaTaBa).

It is worth mentioning that the vowel letters in Arabic are one of these letters {ي,ا,و} - {YaE, AaLiF, WaW}.

Regular verbs, in their turn, are divided into three sub-classes, as Fig. 1 shows, [sound –سَالِم–SaLiM], [Duplicated – مُضَعَف – MoDaAaEF] and [hamzated –مَهْمُوز - MaHMoZ].

Sound Verbs. Their radical letters are free of hamzaletters(أ-ء-ئ-ؤ), and none of their radicals are identical. In its turn, this type of verbs is divided into three sub classes. The reason of this division is due to the inflections forms. All verbs that endwith the letter [n - ن] are inflected in a similar way to those that end with the letter [t - ت] except the first singular person and the second plural person, to which we have to add *shedda*ӧdiacritic.

Duplicated Verbs. They contain all verbs that their second and third root letters are identical. For example, the root letters of the verb (to count -عَّ- AaDa) are (ADD-عدد); the diacritic sign ӧ refers to the duplication of any letteron which it is placed. This category of verbs is divided into two sub-classes: category * (all verbs that their root letters are free of hamza) like the verb (to count – AaDDa- عَّ), its root letters are: [عد]

free of hamza and the second category is hamzated (verbs that contain hamzaletter in one of their three radicals). These two categories are divided, in their turn, three sub-categories [7].

Hamzated Verbs. The category of the verbs that contain hamzaletter in any of its root letters like the verb (to read - قَرَأَ -KaRaAa), is divided into tree sub-classes, as it is shown in Fig. 1.

3.3   Inflectional and Derivational Forms of Arabic Verbs

Arabic derivations are derived from the combination of a specific pattern and a given root. For example, the matching process between the root (LAAB-لعب) and the pattern (FaAiLon- فَاعِلٌ), that serves to generate the active participle form, gives the noun (player – LaAiBon-لَاعِبّ). The same thing is applied to the rest of the derivation forms of all Arabic verbs.

Also, the matching process gives the inflections of any Arabic verb as we have explained before, or by applying several morphophonological alternations to this combination.

Arabic verbs inflect for the following grammatical categories:

o   *Subject agreement*: in person (1ˢᵗ, 2ⁿᵈ and 3ʳᵈ), number (singular, dual, and plural) and gender (masculine and feminine)
o   *Tense / aspect*: (perfect الَمَاضِي-imperfect المُضَارِع)
o   *Mood*: (indicative المَرفُوع, subjunctive المَنْصُوب , jussive المَجْزُوم, long energetic المُؤَكّد الثَّقِيْل, imperative الأَمْر, imperative of long energetic الأَمْرالمُؤَكّد) and
o   *Voice* (active المَبْنِي لِلْمَعْلُوم and passive المَبْنِي لِلمَجْهُول) [6].

Along with the root and pattern, we have adopted the previous grammatical categories as the text annotation. The result of the text analysis appears as annotations that give all information of the grammatical category, root and pattern of all verbs in the text.

4   Practical Study

We are going to review the previous theoretical study using NooJ platform, in order to formalize a comprehensive model of Arabic verbs, by building a dictionary of verbs that is based on the root and pattern approach. This dictionary will use our lexical grammars to generate all possible verb inflections and derivations.

NooJ is a linguistic developmental environment, which can analyze texts of several million words in real time. It includes tools to construct, test and maintain large coverage of lexical resources, as well as morphological and syntactic grammars. Dictionaries and grammars are applied to texts in order to locate morphological, lexicological and syntactic patterns, remove ambiguities, and tag simple and compound words [2].

As we have mentioned before, the following words: ([to study –DaRaSa - دَرَسَ] [a teacher - moDaRiS- مُدْرِس], [a school – maDRaSa - مَدْرَسَة], [a study - DiRaSa- دِرَاسَة], [a lesson – DaRS - دَرْس] and [to teach – DaRraSa - دَرّسَ] are formed as entries and they share the same root. They share also the same concept. Even though they have different patterns. We can retrieve all of these words with their inflectional and derivational forms within a text, using the root that they share. We can also apply a search using pattern to extract all words that share the same concept (as we have explained before), while EL-DiCardictionary cannot retrieve concepts, because it is based on lemma and their entries are separated.

Figure 2 shows the result of the search operation in EL-DiCardictionary using lemma [to write – كَتَبَ] as search entry, on a text that contains these words: ([Library - maKTaBa-مَكْتَبَة], [Writer-KaTiB-كَاتِب], [Write a letter-KaTaBa- كَاتَبَ], [To copy - iKTataBa-إكْتَتَبَ] and [Desk-maKTaB-مَكْتَب]), while Fig. 3 shows the same operation's result on the same text using the root as a search entry. As we can observe the first search operation in Fig. 2 that is based on lemma [to write – كَتَبَ] will retrieve only the active participle because it is formed as a derived form of this entry in EL-DiCar dictionary. Thus, this dictionary provides only inflectional and derivational forms of a given lemma.

Furthermore, unlike

The difference between our provided root and pattern approach and lemma approach, used by EL-DiCar dictionary is very clear by applying a search using lemma [to write – كَتَبَ] on a text that contains words which share the same root: ([Library - maKTaBa-مَكْتَبَة], [Writer-KaTiB-كَاتِب], [Write a letter-KaTaBa- كَاتَبَ], [To copy - iKTataBa-إكْتَتَبَ] and [Desk-maKTaB-مَكْتَب]) returned only the active participle that is formed as a derived form of the entry (to write – KaTaBa- كَتَبَ) in EL-DiCar dictionary. While the rest of the words are not retrieved as they are not linked with the lemma (to write – KaTaBa- كَتَبَ).

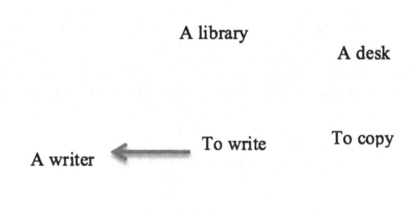

Fig. 2. The result of lemma search in EL-DiCardictionary.

Figure 3 shows the result of our dictionary that is based on root and pattern approach. All words that share the same root will be retrieved using root search.

Now, we move to explain the internal structure of EL-DiCar dictionary. To clarify In lemma based dictionary, the entry that represents a verb takes the form دَرَّسَ, V + Tr + FLX = V_darrasa + DRV = D_darrasa:FLXDRV:(دَرَسَ –DaRaSa- to teach) is an entry; V: Verb, Tr: Transitive, V_darrasa: the inflectional paradigm that gives 122

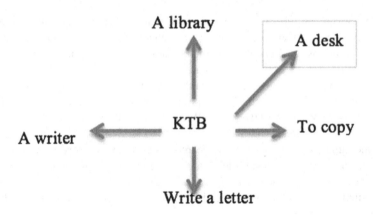

Fig. 3. The result of root search in our dictionary.

inflectional transformations, for example: [I thought – DaRaSTo - دَرَسْتُ] and [you thought – DaRaSTa - دَرَسْتَ], and DRV = D_darrasa: derivational paradigm that gives the active participles (مُدَرِّس MoDaRiSon   teacher) and the passive participles (MaDRouS – مَدْرُوسّ – has been taught) of the entry, and other plural forms [7]. As we have mentioned before, both the active and the passive participles are derived from the same entry. Conducting a search using this entry will retrieve both of them if they exist within the text.

The following words: دِرَاسَة, N + FLX = N_drassa2 (study - DiRaSa-دِرَاسَة) and مَدْرَسَةّ, N + FLX = N_mdrassa, (a school – maDRrSa - مَدْرَسَة) share the same root. Making a search using (DiRaSa-دِرَاسَة - a study) or any of their inflectional or derivational formswill neither give (a school – maDRrSa - مَدْرَسَة) nor their derivations and inflections if they exist within the same text.

Unlike EL-DiCardictionary, in our dictionary, we have added more morphological features: root and pattern. For example our dictionary gives the following forms of entries:

- درس,دَرَسَ,V+Tr+درس+فَعَلَ+FLX=flx1+DRV=drv1:flx2.
- درس,مَدْرَسَة,N+درس+مَفْعَلَة+FLX=N_mdrassa+DRV=drv2:flx3.
- درس.root of: FLX=N_drassa+DRV=drv3:flx4 فِعَالَة+درس+N+,درس,دِرَاسَة the entries, [(فَعَلَ-FaAaLa), (مَفْعَلَة-maFAaLa),(فِعَالَة-FiAaLa)]: patterns of the entries, FLX and DRV the inflectional and the derivational paradigms, that serve to generate the inflections and the derivations of the entries.

The difference between the internal structure of EL-DiCardictionary and our dictionary lies in the adopted approach as we have previously explained.

Furthermore, unlike our dictionary EL-DiCar dictionary cannot make a root and pattern search. This is because EL-DiCar dictionary lacks the morphological features (root and pattern) that we are going to use to make an advanced search.

Making a search using the root [DRS-درس] in our dictionary, will retrieve all entries with their inflectional and derivational forms that share the same root like: ([to study – DaRaSa - دَرَسَ] – [a teacher - moDaRiS- مُدَرِس] – [a school – maDRaSa - مَدْرَسَة] - [a study - DiRaSa- دِرَاسَة] - [a lesson – DaRS - دَرْس] – [to teach – DaRraSa - دَرَّسَ]).

Figure 4 contains a text that we are going to analyze in NooJ platform, using our dictionary that is based on root and pattern approach, and linked to our lexical grammars that generate the possible inflectional and derivational verb forms.

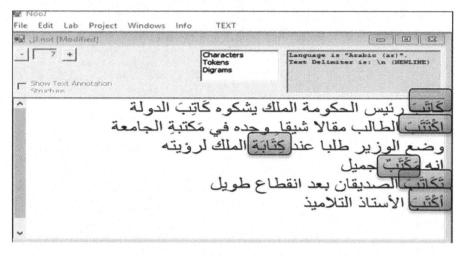

Fig. 4. Text to be analyzed in NooJ platform using EL-DiCar dictionary

The text: The prime minister *wrote* to the king to complain about his State *Secretary*, A student *copies* an interesting article that he found at the University *Library*; the minister submits a demand to the king *secretary* asking for an appointment to meet the king; it is a beautiful desk; two friends *wrote* to each other after a long time; The teacher *dictates* to his students. Figure 5 shows the "locate pattern" or the search wizard in NooJ platform [8]; the lemma (to write – KaTaBa- كَتَبَ) should return all inflectional and derivational forms of this entry. Note that NooJ Platform allows us to make a search with full, semi or even without diacritics [8].

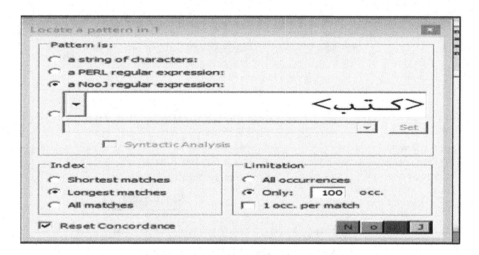

Fig. 5. Lemma search in NooJ platform using DiCar dictionary-EL

In the result of Fig. 6, only the derivational form (active participant) of the previous entry is retrieved, although the text contains seven words that share the same root [KTB-كتب].

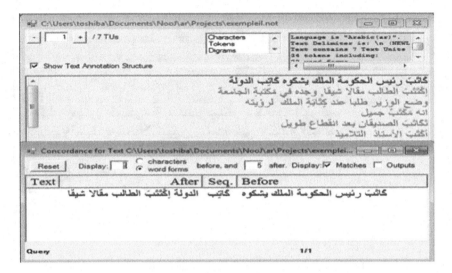

Fig. 6. Lemma search result in NooJ platform using dictionary DiCar-EL

Figure 7 shows the search operation using the root (KTB-كتب) on the previous text, using our dictionary:

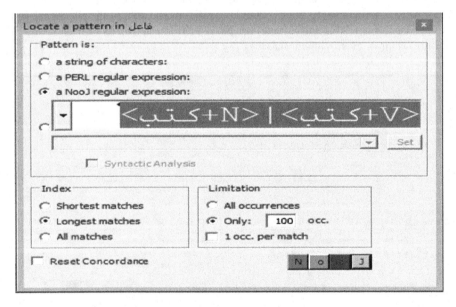

Fig. 7. Root search in NooJ platform

The query above retrieves all verbs and nouns that have the root KTB. The result is as follows:

All entries with their inflectional and derivational forms that share the root KTB have been retrieved, as it is shown in Fig. 8.

Fig. 8. Root search result in NooJ platform using our dictionary.

While patterns care about meaning, we can also apply a pattern search in a text, to retrieve all words that share the same concept. For instance, the pattern [FiAaLa – فِعَالَة] that refers to a craft will retrieve all crafts that occur in the text that Fig. 9 shows:

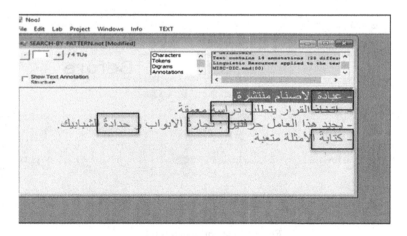

Fig. 9. Text to be analyzed in NooJ.

- The *worship* of idols is widespread.
- Decision-making requires an in-depth *study*.
- The worker mastered two craft: wood *carpentry* and *wild fences*.
- *Writing* examples is exhausting.

As we can see, the text contains five words that share the same pattern [FiAaLa – فِعَالَة]. Making a search using this pattern will retrieve all crafts that occur in the text. Figure 10 shows the search operation using the previous pattern.

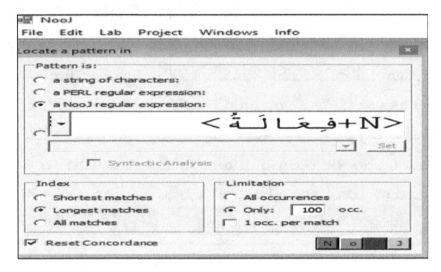

Fig. 10. Pattern search.

The result is as follows (Fig. 11):

Fig. 11. Pattern search result.

We can retrieve also all nouns that refer to an actor using the pattern [FaAiL‫فَاعِل‬] or all nouns that refer to an action or a place using their patterns.

Text annotations take the form shown in Fig. 12. The text contains 213 inflectional forms of the verb (to store – KhaZaNa - ‫خَزَنَ‬); its root is [KHZN-‫خزن‬], voice = passive, tense = time = I, while I means perfect, subject agreement = 1st person.

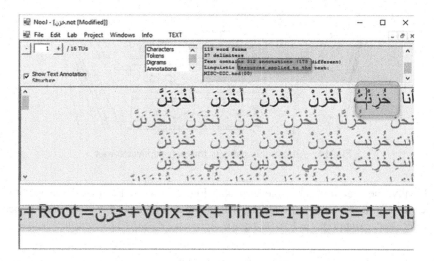

Fig. 12. Text annotation form.

Lexical grammar takes the same hierarchy of the verb classification, in NooJ platform; Fig. 13 gives an example of a lexical grammar of the regular category. Yellow color means that the case includes is a sub-graph.

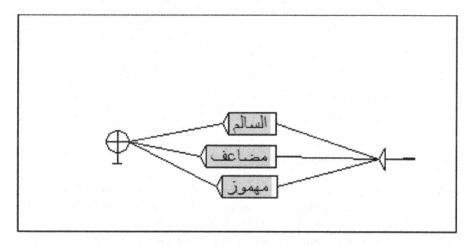

Fig. 13. A sub graph of the regular category

As we have mentioned before, this category contains tree sub categories: sound-duplicated-hamzated, (see Sect. 3.2).

An example of a sub-graph for the duplicated category, as Fig. 14 shows: (FaaAaLa-فَاعَلَ); (taFaAaLA-تَفَاعَلَ) and (inFaAaLa- إنْفَعَلَ): several sub-graphs that generate all possible inflectional and derivational forms of a given duplicated root;

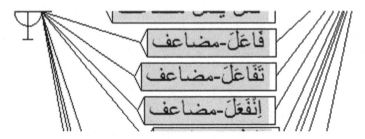

Fig. 14. All possible patterns for each duplicated root

Finally, the grammar takes the form that Fig. 15 shows. Each of these operators serves to reform the root in the inflectional or the derivational form. (For more operators see NooJ Manuel).

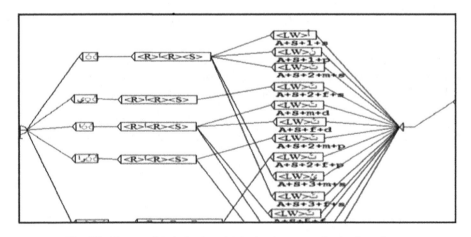

Fig. 15. The possible inflections in the jussive case of a duplicated root

5 Conclusion and Perspectives

Our dictionary that is based on root and pattern approach contains 14500 entries, generated from 295 verbs models. These models cover all Arabic verb categories, each entry contains all possible inflectional and derivational forms, that are generated using our grammar. The adopted approach allows us to extract all words within a text that share the same root. We can also extract concepts using a pattern search. The result of the linguistic analysis of the text appears as annotations that give all morphological details about the analyzed text words. In future work we will cover the other Arabic words like nouns and adjectives. Also adding Arabic morphophonological alternation as new grammars.

References

1. Mourchid, M.: Génération morphologique et applications. Thèse de Spécialité de 3ème Cycle, Université Mohammed V, Juillet (1999)
2. Slim, M.: Standard Arabic formalization and linguistic platform for its analysis. In: The Challenge of Arabic for NLP/MT, LASELDI, Franche-Comté University, France (2006)
3. Nabil, A.: Arabic language and computer. Ta'areeb (1988). (in Arabic)
4. Nooj   association.   http://www.nooj-association.org/index.php?option=com_k2&view= item&id=2:arabic-resource&Itemid=611
5. Karin, C.: A Reference Grammar of Modern Standard Arabic. 2nd edn. Cambridge University Press, New York (2005)
6. Antwan, D.: A dictionary of universal Arabic grammar. Lebanon liberary, dar el nacher wa almaajem (1999). (in Arabic)
7. Slim, M.: Analyse morpho-syntaxique automatique et reconnaissance des entites nommees en arabe standard. Specialty thesis of 3rd round, Doctoral thesis, LASELDI, Franche-Comté University, France (2008)
8. Silberztein, M.: Nooj Manual. www.nooj-association.org

Syntactic Analysis

Arabic NooJ Parser: Nominal Sentence Case

Nadia Ghezaiel Hammouda[1] and Kais Haddar[2(✉)]

[1] Miracl Laboratory, Higher Institute of Computer and Communication
Technologies of Hammam Sousse, Sousse, Tunisia
ghezaielnadia.ing@gmail.com
[2] Miracl Laboratory, Faculty of Sciences of Sfax,
University of Sfax, Sfax, Tunisia
kais.haddar@yahoo.fr

Abstract. Parsing Arabic corpora is an important task aiming to understand Arabic language, enrich and enhance the electronic resources, and increase the efficiency of natural language applications like translation or the recognition. In this paper, we propose a parsing approach for Arabic sentences especially for nominal ones. To do this, we first study the typology of the Arabic nominal sentence. Then, we develop a set of rules generating different nominal sentences. After that, we present our parsing approach based on transducers and on our tag set. In addition, we transform recursive graph of transducers into transducer cascade to reduce the complexity. Finally, we present the implementation and experimentation of our approach in NooJ platform. The obtained results are satisfactory.

Keywords: Arabic sentence · Recursive graph · Topic · Attribute
NooJ transducer

1 Introduction

Parsing Arabic corpora is an important task aiming to understand Arabic language, enrich and enhance the electronic resources, and increase the efficiency of naturel language applications like translation or the recognition. Arabic is considered as one of the difficult language to analyze due to its morphological, syntactic, phonetic and phonological characteristics. There are two types of sentences in Arabic: the verbal sentence and the nominal sentence.

There are different forms of the nominal sentence that can interact with verbal sentences. The formalization of rules requires much effort to guarantee several qualities like efficiency, robustness and extensibility. Transducers have proved their usefulness in a wide variety of applications in NLP [16]. Transducer cascades made possible to carry out robust and highly precise syntactic analysis on different corpora.

Transforming recursive graph of transducers into transducer cascade is very interesting. The transformation is a difficult task due to the difference between the application levels in every path and the interaction of the linguistic phenomena. For the cascade, the order of transducers should respect different constraints, which are deduced from observations done on Arabic corpora.

S. Mbarki et al. (Eds.): NooJ 2017, CCIS 811, pp. 69–80, 2018.
https://doi.org/10.1007/978-3-319-73420-0_6

Our objective is to construct an Arabic parser implemented in NooJ. To do this, we will study, essentially, the Arabic nominal sentences but also other sentence forms. Then we will establish a set of rules recognizing nominal sentences that can be generalized to treat any sentence type. Finally, we will implement the transducer cascade in NooJ.

In this paper, we begin by stating the different approaches, which allow the parsing and annotation of Arabic corpora. Then, we perform a study about the forms of Arabic nominal sentences. Next, we establish syntactic rules transformed in transducers. In addition, we implement and test all these rules in the NooJ platform respecting the cascade notion. Finally, we provide a concise conclusion and we give some future perspectives.

2 Previous Work

Many works aim to analyze Arabic corpora with different approaches: rule-based, statistical or hybrid approach. In [1], the authors have proposed a method for Arabic lexical disambiguation based on the hybrid approach. In [2], the author has developed a morphological syntactic analyzer for the Arabic language within Lexical Functional Grammar formalism. The developed parser is based on a cascade of finite-state transducers and a set of syntactic rules specified in Xerox Linguistics Environment. Also, in [3], the authors have proposed a rule-based approach for tagging non-vocalized Arabic words. In [4], the authors have designed an automatic tagging system by adding the part-of-speech tag in the Arabic text. In addition, in [5], the authors have presented an Arabic parser for Arabic nominal sentences. In this work, the HPSG formalism is used.

In addition, there are many other statistical and hybrid works. In [6], the proposed method of parsing dealt with the 'alif-nûn' sequence in a given sentence. This method is based on the context-sensitive linguistic analysis to select the correct sense of the word in a given sentence without doing a deep morpho-syntactic analysis.

Besides, in the last decades, many researchers have worked on systems, which aim to disambiguate Modern Standard Arabic. Among those systems, we mention MADA and TOKAN systems [7]. They are two complementary systems for the Arabic morphological analysis and disambiguation process. Their applications include high-accuracy part-of-speech tagging, discretization, lemmatization, disambiguation, stemming and glossing. In [8], the system AMIRA developed at Stanford University includes a tokenizer, a part of speech tagger (POS) and a Base Phrase Chunker (BPC). The model used by AMIRA is a supervised learning machine with no explicit dependence on knowledge of deep morphology. Concerning the finite state tools, we find the Xerox parser [9], which is based on finite state technology, tools (e.g. xfst, twolc, lexc,) for NLP. These tools have been used to develop the morphological analysis, tokenization, and shallow parsing of a wide variety of natural languages.

Moreover, there are several parsing works performed with the NooJ platform. In [10], the authors proposed a method to identify all possible syntactic representations of the Arabic relative sentences. The authors explain the different forms of relative clauses and the interaction of relatives with other linguistic phenomena such as ellipsis and coordination. We can cite also the work described in [11] to analyze the Arabic broken plural. This work is based on a set of morphological grammars used for the detection of the broken plural in Arabic texts. Arabic broken plural analysis can facilitate the parsing because we can distinguish between different types of nouns.

3 Arabic Lexical Ambiguity

The Arabic language is written and read from right to left. The alphabet has 28 consonants, adopting different spellings according to their position (at the beginning, middle or end of a lexical unit). Arabic token is written with consonants and vowels. The vowels are added above or below the letters. The presence of vowels allows us to understand text and disambiguate different words. In Arabic, the word should respect a well-defined type hierarchy. Indeed, a word can be either a verb or a name or a particle. Each type itself is detailed in several subtypes. Thus, any specific linguistic information to the Arabic language should be represented through this hierarchy. Before beginning our study of lexical ambiguity, we give an overview about some specificities of Arabic language. Indeed, the Arabic sentence is characterized by a great variability in the order of its words. In general, in Arabic, we put at the beginning of the sentence the word (noun or verb) on which we want to attract the attention and at the end the richest term to keep the meaning of the sentence. This variability in the order of words causes artificial syntactic ambiguities. So in the grammar, we should give all possible combinations of inversion rules for the word order in the sentence. Note that the Arabic sentence can be either verbal or nominal.

Arabic lexical ambiguity has several causes, but we focus mainly on five of them.

Unvocalization: It can cause lexical ambiguities because a word in Arabic language can be read differently in a sentence, depending on its context. For example, the word *kaataba* can refer to the noun (the writer), or the verb *to write* in English.

Emphasis sign (Shadda ٥ّ): In Arabic, the emphasis sign Shadda is equivalent to writing the same letter twice. The insertion of Shadda can change the meaning of the word. For example, the word *darasa* means *lessons* (noun) while *darrasa* means *he taught* (verb).

Hamza sign: The presence of Hamza sign (hamzah) reduces ambiguity. If we add the Hamza to a word then the number of ambiguities decreases. As an example, the word Faas can be a city or an ax.

Agglutination: In Arabic, particles, prepositions, pronouns, can be attached to adjectives, nouns, verbs and particles. This characteristic can generate many types of lexical ambiguity. For example, the letter *faa'* in the word *fa-slun* (season) is part of the root while in the word *fasala* (then he prayed) is a prefix.

Compound words: Lexical ambiguity sometimes derives from compound words. For example, the compound noun "الحاسوب المحمول" hassub mahmul can be interpreted as a laptop or a portable pc.

4   Typology of Nominal Sentence

As we have mentioned, the Arabic language has two types of sentences: the nominal sentence and the verbal one. In the following section, we will present the typology of the nominal sentence. The nominal sentence is any sentence beginning with a noun and can contain a verbal sentence as a component. Also, each nominal sentence is composed of a topic (*Mubtada '*) and an attribute (*Khabar*) and the attribute is compatible with the topic in gender and number. From this definition, we can identify several types of the Arabic nominal sentence.

4.1   Structure of Nominal Sentence

The topic and the attribute can be presented in many forms. In what follows, we detail these forms. In our study, we concluded that the topic could have many forms. It can be a single word, a phrase or a sentence.

(a) The case of a single noun: In this case, the topic can be a proper noun (name of person, geographical name, etc.) or a common noun. Also, it can be a personal pronoun, a demonstrative pronoun or an interrogative pronoun. Examples from (1) to (4) illustrate this case.

(1) مريم جميلة
Mariam [is] beautiful
(2) الطاولة مستديرة
The table [is] round
(3) أنت جميلة
You are beautiful
(4) هذا صديقي
This is my friend

(b) The case of a nominal phrase: In this case, the topic can be a phrase of annexation, an adjectival phrase, a relative clause or a phrase of conjunction. Also, each one of those phrases can be recursive or contain one of the other. To illustrate this case we present the following examples:

(6) باب الحديقة جميل
The door of the garden is beautiful
(7) باب حديقة المنزل جميل
The door of the garden's home is beautiful

The example (6) presents a phrase of annexation which is composed of an indefinite noun (باب) and a definite noun (الحديقة), but the example (7) presents a recursive phrase which contains another phrase of annexation (حديقة المنزل) and (حديقته).

The attribute is manifested in several forms. It can be a unique word, a phrase or a verbal sentence.

(a) The case of a unique word: In this case, the attribute can be a noun, a personal pronoun, an intransitive verb, or an adjective. We illustrate this case by examples (8) and (9).

(8) العلم نور
Knowledge is the light
(9) الولد يضحك
The boy laughs

(b) The case of a phrase: generally the attribute is in the form of a phrase. It can be a nominal phrase (example (10)), a prepositional phrase (example (11)) or a relative phrase (example (12)).

(10) الهدهد طائر جميل
Hoopoe is a beautiful bird
(11) الأولاد في المدرسة
Boys are at school
(12) محمد الذي نجح في الإمتحان لمثابرته
Mohammed who passed the exam, due to his perseverance

(c) The case of a sentence: the attribute can be a verbal sentence or a nominal sentence. To illustrate this case, we present the following examples:

(13) المديرة تمنح التلاميذ المميزين الجوائز
The director gives presents to distinguished students
(14) الله هو النور الأعظم
Allah is the greatest light

In the example of (13), the attribute is a verbal sentence. On the other hand, the attribute of example (14) is a nominal sentence.

4.2 Other Types of Nominal Sentence

In Arabic, the nominal sentence can be introduced by particles such as the particle *Inna* or defective verbs such as the verb *Kaana*. The insertion of defective verbs or particles in a nominal sentence can change the joint of the topic and the attribute. In fact, the particle *Inna* accepts a subject and a predicate through dependencies called *Ism inna* (اسم إن) and *khabar inna* (خبر إن). The subject *ism inna* is always in the accusative case *manṣūb* (منصوب) and the predicate *khabar inna* is always in the nominative case *marfū*. The example of sentence (15) uses the particle *Inna* the topic becomes accusative but the attribute stays nominative. The same sentence of (16) without the particle *Inna* keeps its characteristics.

(15) إنَّ الإمتحانَ صعبٌ
The exam is difficult
(16) الإمتحانُ صعبٌ
The exam is difficult

5   Formalization of Lexical Rules

We carried out a linguistic study, which allowed us to identify lexical rules and resolve several forms of ambiguity. The identified rules were classified through the mechanism of subcategorization for verbs, nouns and particles [12].

Particles can be subdivided into three categories: particles acting on nouns, particles acting on verbs and particles acting on both nouns and verbs. There are Arabic particles which must be followed by a noun like prepositions and particles of restriction {مِنْ، إلى، عن، على، في، ب، ل، ك، حتّى، رُبَّ، واو القسم، ت، حاشا، خلا،عدا}.

Particles can also be followed by a verb, like subjunctive particles, apocopate particles, prohibition particles. As an example, if we find a subjunctive particle like {لن/ كي/ حتى/ لام التعليل/ إذن/ فاء السببية/ وأو المعية/ لام الجحود/أن}, then, it should be followed by a verb. A noun or a verb can follow some particles. To solve this ambiguity, we studied the context of the sentence.

We can apply the principle of sub-categorization to resolve the ambiguity related to verbs. We based essentially on the transitivity feature of verbs. In Arabic, a verb can be intransitive, transitive, di-transitive and tri-transitive. Either transitive or intransitive verbs can be transformed to transitive verbs with prepositions. The mechanism of transitivity is explained by the above sentences. Note that these examples respect the VSO order.

We can also apply the principle of sub-categorization to resolve the ambiguity linked to nouns. We based essentially on successors feature of nouns. In Arabic, a noun can be defined with 'ال' *'alifLam'* or be indefinite. Each one of these types has its followers. The defined noun can be followed by a noun phrase (NP), a defined adjectival phrase (AP), a prepositional phrase (PP), a relative phrase (RP) or an empty set (∅). Besides, the non-defined noun can have the same followers but the AP should be non-defined. Note that, the nominative, accusative or genitive criteria will be inherited by the nominal group.

To implement our rules, we use the linguistic platform NooJ which is a linguistic environment to build and manage electronic dictionaries and grammars with wide coverage and to formalize various language levels: spelling, inflectional and derivational morphology, lexicon of simple words, compound words and idioms, local syntax and disambiguation, structural and transformational syntax, semantics and ontologies. Also, formalized descriptions can then be used to process and analyze texts and large corpus.

6 Proposed Method

Our proposed approach of analysis consists of two main phases: the segmentation and the parsing.

The segmentation phase [13] consists of the identification of sentences based on punctuation signs. Each identified sentence is delimited by an XML tag. As an output of this phase, we obtain an XML document for the corpus, and it will be the input for the pre-processing phase. The second phase consists of the agglutination's resolution using morphological grammars. As an output of this phase, we obtain a Text Annotation Structure (TAS) containing all possible annotations for corpus's sentences. The obtained TAS is the input of the third phase. Then, we identify the appropriate lexical category of each word in the sentence to construct different sentence phrases. This identification is based on several syntactic grammars specified with NooJ transducers. Transducer's applications respect a certain priority from the most evident and intuitive transducer until arriving at the least one (Fig. 1). The output of the parsing phase will be a disambiguated TAS containing right paths and right annotations. Note that we used a high granularity's level for lexical categories. This distinction between nominative, accusative and genitive modes for nouns can resolve the absence of vocalization. Another remark, we have tested two methods to analyze Arabic nominal sentences.

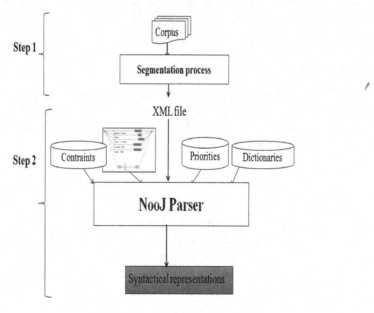

Fig. 1. Proposed method

7   Implementation

The extracted rules have been implemented in the NooJ platform [14]. In fact, the process of parsing is based on the set of the developed NooJ transducers and a tag set that is indicated in the following (Table 1).

Table 1. Used tag set

NN	Indefinite Nominative Noun u
NTN	Indefinite Nominative Noun un
NND	Definite Nominative Noun u
NA	Indefinite Accusative Noun a
NTA	Indefinite Accusative Noun an
NAD	Definite Accusative Noun a
NG	Indefinite Genitive Noun i
NTG	Indefinite Genitive Noun in
NGD	Definite Genitive Noun i

In this part, we will explain different stages in our cascade approach by giving an idea about the recursive approach.

7.1   Segmentation Phase

The implementation of the segmentation phase is based on a set of developed transducers in the NooJ linguistic platform. This set contains 9 graphs representing contextual rules. The main transducer adds an XML tags <S> to delimit the frontiers of a sentence.

7.2   Preprocessing Phase

The implementation of the preprocessing phase is based on a set of morphological grammars and dictionaries [15] existing in the NooJ linguistic platform (Table 2). This implementation resolves all forms of agglutination. The outputs contain all possible lexical categories of each word in sentences.

Table 2. Table summarizing morphological grammars

Morphological grammars	Numbers
Verb inflected form patterns	113
Inflected relative pronoun patterns	8
Broken plural patterns	10
Agglutination's grammars	3

7.3 Analysis Phase with Recurisve Graphs

Figure 2 illustrates the NooJ implementation of rules for nominal sentences.

In fact, Fig. 2 shows different forms of topics and attributes. A nominal sentence can be formed by a nominative topic followed by a nominative attribute. Also, we can find the modal verb "KANA" followed by a nominative topic and an accusative attribute. In addition, we find the modal verb "INNA" followed by an accusative topic and a nominal attribute. In the case of a simple nominal phrase, the topic and the attribute should have the same joint. They respect the nominative form.

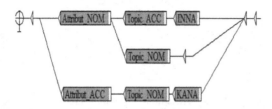

Fig. 2. Transducer representing a lexical rule for a nominal sentence

Figure 3 shows that the topic can be a unique word, a unique noun phrase or recursive one. Note that the subgraph PP represents the different forms of the prepositional phrases and the subgraph NP_NOM represents the noun phrase. For a nominative attribute, the appropriate transducer is given in Fig. 4.

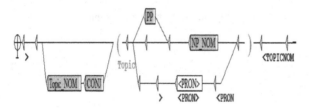

Fig. 3. Transducer representing a rule for a nominative topic

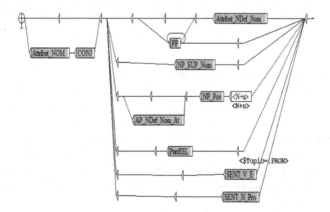

Fig. 4. Transducer representing a lexical rule for a nominative attribute

7.4   Cascade for Parsing

A separated transducer implements each nominal sentence component. In what follows, some transducers respecting the proposed approach are given.

Figures 5, 6 and 7 show how our cascade works and show that the different transducers use automatically the calculated output.

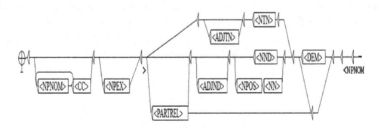

Fig. 5. Transducer for nominative NP

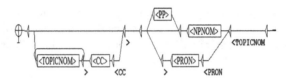

Fig. 6. Transducer for a topic

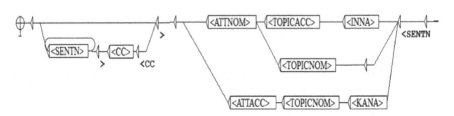

Fig. 7. Transducer for nominal sentence

8   Experimentation and Evaluation

To experiment our approach, we implemented our cascade of transducers in NooJ platform. Then, we compared the cascade with recursive transducers in the case of the nominal sentence. In fact, fixing the call order of transducers was inspired by our study. To be more specific, the idea consists of starting with phrases until gathering the sentence entirely: Particles → Phrases → Sentences. The implemented syntactic transducer cascade contains in total 50 graphs called in a fixed order. This is illustrated in the following (Fig. 8).

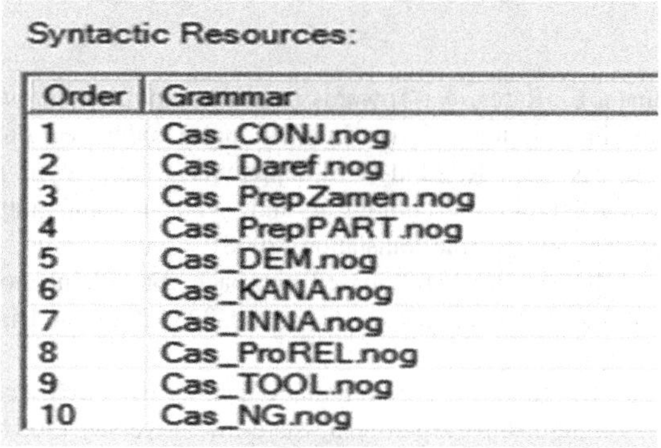

Syntactic Resources:

Order	Grammar
1	Cas_CONJ.nog
2	Cas_Daref.nog
3	Cas_PrepZamen.nog
4	Cas_PrepPART.nog
5	Cas_DEM.nog
6	Cas_KANA.nog
7	Cas_INNA.nog
8	Cas_ProREL.nog
9	Cas_TOOL.nog
10	Cas_NG.nog

Fig. 8. Syntactic resources

To evaluate our prototype, we calculate also the precision, the recall and the F-measure for two approaches using respectively recursive graphs and cascade, as illustrated in Tables 3 and 4.

Table 3. Table summarizing the precision and recall measures for recursive graphs

Corpus	Precision	Recall	F-measure
5900 sentences	0.6	0.7	0.64

Table 4. Table summarizing the precision and recall measures for the proposed cascade

Corpus	Precision	Recall	F-measure
5900 sentences	0.74	0.84	0.77

The obtained values of these measures are interesting and show that a cascade method is better than a recursive one. This result can be improved by adding other rules and heuristics.

9   Conclusion

In this paper, we have proposed a parsing method dealing with the Arabic nominal sentences. This method is based on a set of transducers and a high level of granularity. This method is implemented in the NooJ platform and used a cascade instead of recursive graph. The elaborated parser can annotate the Arabic corpora. So, we did a study on different forms of Arabic nominal sentences. This study allowed us to establish a set of rules for parsing Arabic nominal sentences. The established rules are specified with NooJ transducers. The proposed cascade of transducers reduces the parsing complexity. Thus, an experiment is performed on a set of nominal sentences, mainly from stories. The obtained results are satisfactory, which is proved by the calculated measures. Concerning the future works, we want to enrich our linguistic resources by improving our dictionaries and transducers.

References

1. Shaalan, K., Othman, E., Rafea, A.: Towards resolving ambiguity in understanding Arabic sentence. In: The Proceedings of the International Conference on Arabic Language Resources and Tools, NEMLAR, 22nd–23rd September, Cairo, Egypt, pp. 118–122 (2004)
2. Attia, M.: Handling Arabic morphological and syntactic ambiguity within the LFG framework with a view to machine translation (2008)
3. Al-Taani, A.T., Al-Rub, S.A.: A rule-based approach for tagging non-vocalized Arabic words. Int. Arab J. Inf. Technol. (IAJIT) **6**(3), 320–328 (2009). 4 Diagrams, 5 Charts, 1 Graph
4. Diab, M., Hacioglu, K., Jurafsky, D.: Automatic tagging of Arabic text: from raw text to base phrase chunks. Linguistics Department, Stanford University (2004)
5. Hadddar, K., Abdelkarim, A., Ben Hamadou, A.: Étude et analyse de la phrase nominale arabe en HPSG. In: TALN 2006 (2006)
6. Dichy, J., Alrahabi, M.: Levée d'ambiguité par la methode d'exploration contextuelle: la sequence 'alif-nûn' en arabic. In: Second International Conference (SIIE) (2009)
7. Habash, N., Rambow, O., Roth, R.: MADA+TOKAN: a toolkit for Arabic tokenization, diacritization, morphological disambiguation, POS tagging, stemming ald lemmatization. In: Proceedings of the 2nd International Conference on Arabic Language Resources and Tools (MEDAR), Cairo, Egypt (2009)
8. Diab, M.: Second generation tools (AMIRA 2.0): fast and robust tokenization, POS tagging, and base phrase chunking. In: MEDAR 2nd International Conference on Arabic Language Resources and Tools, April, Cairo, Egypt (2009)
9. Beesley, K.: Finite-state morphological analysis and generation of Arabic at xerox research: status and plans. In: ACL/EACL 2001, 6th July, Toulouse, France (2001)
10. Zalila, I., Haddar, K.: Construction of an HPSG grammar for the Arabic relative sentences. In: Natural Language Processing, RANLP 2011, 12–14 September 2011, Hissar, Bulgaria (2011)
11. Ellouze, S., Haddar, K., Abdelhamid, A.: Etude et analyse du pluriel brisé arabe avec la plateforme NooJ (2009)
12. Hammouda, N.G., Haddar, K.: Toward the resolution of Arabic lexical ambiguities with transduction on text's automaton. In: CICLing (2015)
13. Hammouda, N.G., Haddar, K.: Integration of a segmentation tool for Arabic corpora in NooJ platform to build an automatic annotation tool. In: Will appears in NooJ (2016)
14. Silberztein, M.: Disambiguation tools for NooJ. In: Proceedings of the 2008 International NooJ Conference, pp. 158–171. Cambridge Scholars Publishing, Newcastle (2010)
15. Mesfar, S.: Analyse morpho-syntaxique automatique et reconnaissance des entités nommées en arabe standard. University of Franche Comté, p. 235 (2008). Thesis
16. Gross, M.: The Construction of Local Grammar. Finite-State Language Processing, pp. 329–354. MIT Press, England (1997)

The Parsing of Simple Arabic Verbal Sentences Using NooJ Platform

Said Bourahma$^{(\boxtimes)}$, Mohammed Mourchid, Samir Mbarki,
and Abdelaziz Mouloudi

MISC Laboratory, Faculty of Science, Ibn Tofail University, Kenitra, Morocco
saidbrh@yahoo.fr, mourchidm@hotmail.com,
mbarkisamir@hotmail.com, mouloudi_aziz@hotmail.com

Abstract. In this paper, we present a NooJ parser of simple Arabic verbal sentence. This parser is based on dependency grammar established by the attribution (الإسناد, al-'isnād) concept in the Arabic language. In the first part of this paper, we present a syntactic and semantic classification of Arabic words allowing Arabic sentence parsing. In the second part, we present the shared structure by simple Arabic sentences. Furthermore, we use this structure to implement simple Arabic verbal sentence grammar using NooJ platform. Our parser is applied to the input sentence after two required steps: Morphological analysis and morphological disambiguation. The proposed parser generates possible parse tree(s) of the input sentence, and annotates all sentence components by their grammatical functions. The implemented parser is tested on a selected text; experimental results show its efficacy.

Keywords: Natural Language Processing · Arabic language parser
Syntactic analysis · NooJ linguistic platform

1 Introduction

Arabic language belongs to the Semitic group of languages. In this language, verbs and nouns are often derived from roots constituted with three letters or more using various patterns. Arabic root carries the basic conceptual meaning. Pattern involves lexical, syntactic and semantic information. Arabic textual words are often compound structures, which should syntactically be regarded as phrases rather than single words. Arabic words are divided into three types: Noun, Verb and Particle. Arabic is a relatively free word order language. While the primary word order is verb-subject-object (VSO), Arabic also allows subject-verb-object (SVO), object-verb-subject (OVS), etc.

The printed words, we are reading now, are the perceptible cornerstones of an otherwise invisible grammatical edifice that is automatically reconstructed in our mind. According to many psycholinguists [19], comprehending spoken or written sentences involves building grammatical structures. This activity, which is called syntactic analysis or sentence parsing, includes assigning a word class (part-of-speech) to individual words, combining them into word groups or 'phrases', and establishing syntactic relationships between word groups. All these parsing processes should be in harmony with grammatical rules. Developing a working model of sentence parsing is

S. Mbarki et al. (Eds.): NooJ 2017, CCIS 811, pp. 81–95, 2018.
https://doi.org/10.1007/978-3-319-73420-0_7

impossible without adopting a grammatical formalism and the structure building operations specified in it. Arabic language's syntactic analysis is still in its early stages; most of the researches in Arabic Natural Language Processing (NLP) systems have mainly concentrated on the fields of the morphological analysis. Most works in the Arabic parsing systems have adopted the statistical or hybrid approach [4–7], which cannot give desirable results. Besides, other works have used rule based approach [1–3], which implements separated systems to parsing Arabic sentences in a well-defined order. Extensibility may be difficult in these works to covering many NLP levels to building NLP applications, because the analysis levels must communicate and share results, which must be unified to facilitate this communication. The aim of this paper is to implement a syntactic parser of simple Arabic verbal sentences. Thus, in the first step of our work, we present a syntactic and semantic classification of Arabic words. In the second step, we study the simple Arabic sentence's and simple Arabic verbal sentence's grammatical structure. This parser is based on a dependency grammar thanks to the attribution concept in the Arabic sentence. This grammar is implemented in the NooJ platform. The lack of diacritics in written Arabic texts produces ambiguity at the morphological level. That is why the morphological ambiguity resolution is required in this work before the sentence parsing step.

The rest of this paper is organized as follows: Sect. 2 is dedicated to Arabic lexicon syntactic, and semantic classification, Sect. 3 presents simple Arabic verbal sentence parsing; we have four subsections in this part: simple Arabic sentence study, simple Arabic verbal sentence's grammatical structure, morphological disambiguation and syntactic parsing. In Sect. 4, we present the results and tests of our parser. Section 5 explains the related work. Finally, the last part includes the conclusion and the future work.

2   Lexicon Syntactic Classification

NooJ platform integrates an Arabic dictionary named El-DicAr. This dictionary classifies Arabic words into many classes. Verbs are classified, regarding transitivity feature, into three sub-classes: direct transitive (one direct object), indirect transitive (one indirect object) and intransitive. But transitive verbs are sub-classified according to the number of the accusative form that requires the verb to convey the sentence meaning. Arabic transitive verbs can select one, two or three accusative forms (direct and indirect object(s) in the sentence). Then we cannot use this dictionary to parsing Arabic verbal sentence. In this section, we present a syntactic classification of Arabic words, and we add new proprieties in our dictionary implemented in the work [8] using NooJ platform. This classification includes required features in the syntactic analysis of the Arabic sentence.

2.1   Arabic Lexicon Syntactic Classification

Arabic language linguists classify Arabic words into three main classes: nouns, verbs, and particles [13, 14, 16–18, 20–25]. Each class is in turn divided into sub-classes.

A noun (الاسم) in Arabic is a word that describes a place, person, thing, or an idea, etc. It conveys a lexical meaning, and gives no indication of time. The noun in Arabic language is divided into many sub-classes: complete noun (الإسم التام) like (أسد, lion and قلم, stripe), incomplete noun (الإسم الناقص) like (قرن, ثانية), verbal noun (مصدر)like (قراءة, reading, علم, knowledge), adjective (صفة) such as (قارئ, reader, مكتوب, written, عظيم, great), etc. The sub-class "adjective" has also sub-classes: resembling adjective (صفة مشبهة), "active participle" (اسم الفاعل), "passive participle" (اسم المفعول), etc.

A verb in Arabic language is a word with two features: action and time. A verb refers to an action effected in the past, the present or the future. In fact, the "verbs" class is divided into two main sub-classes: "complete verbs" (الفعل التام, al-fi'lattaam) and "incomplete verbs" (الفعل الناقص, al-fi'l annaqis). We can also classify verbs according to many features. One of them could be the syntactic feature which is considered as an important property of verb transitivity (التعدية). It is used in the syntactic analysis of sentence constructions to determine the number of object elements (arguments), which select the verb, in addition to the subject, to achieve the sentence meaning. Hence, a verb could be (قاصر); it handles only a subject, but its semantic function is a complement so that the verb conveys the sentence meaning. Intransitive (لازم) verb can achieve the sentence meaning only with its subject. Transitive verb (متعدي) requires other syntactic position(s) in addition to the subject to convey the sentence meaning; transitive verbs in Arabic handle from one to three accusative forms. In addition to the syntactic classification, we classify verbs according semantic features such as rationality.

A particle (الحرف) category refers to function words that cannot be considered either as verbs or nouns, and conveys no lexical meaning; it must be linked with another word (noun or verb) to convey meaning. Particles are used to connect words (nouns and verbs together) to make phrases or sentences. They are divided into three categories according to the type of word they can affect. They can either affect a noun, a verb, or both. Particle class includes: prepositions, conjunctions, interrogative particles, exception particles, interjections, etc.

2.2   Classification Implementation

We have already implemented an Arabic dictionary based on root and pattern properties [8]. It is classified based on inflectional and derivational models, number of root letters, morphological features, etc. This dictionary includes all Arabic particles, and 160.000 verbs, nouns and forms (which are also obtained from inflectional and derivational graphs). The implemented dictionary lucks a fine-grained syntactic and semantic classification in its entries. Therefore, we must enrich our dictionary by the properties holding the lexicon syntactic and semantic classification discussed in Sect. 2.1. Table 1 summarizes these properties.

Table 1. Lexicon syntactic classification

Property/ sub-property	Code	Example
- Noun, إسم	N	
- Complete Noun, الإسم التام	N+COM	قلم, pen, qalam
- Incomplete Noun, الإسم الناقص	N+INC	يوم, day , yawm
- Pronoun, الضمير	N+PRO	
- Attached pronoun, ضمير متصل	N+PRO+ATT	ها, her, كُم, yours
- Accusative Pronoun, ضمائر النصب	N+PRO+ACC	إياي, Me, إياك, you
- Nominative Pronoun, ضمائر الرفع	N+PRO+NOM	هو, he, أنت , you
- Demonstrative noun, إسم إشارة	N+DEM	هذا, هذه, this
- Relative noun, إسم موصول	N+REL	التي, الذي, which
- Adjective, الصفة	N+ADJ	
- Verbal Noun, المصدر	N+VRN	دراية, علم
- Instrumental noun, إسم الآلة	N+INS	مضرب, bat
- Interrogative pronoun, إسم الإستفهام	N+ITG	من, ما, what
- Adverb , الظرف	N+ADV	حين, when, مدة , during
- etc.		
- Verb, فعل, fi'l	V	
- Complete Verb, الفعل التام, fi'l taam	V+CMP	
- Transitive 1	V+CMP +TR1	طلب, to request
- Transitive 2	V+ CMP +TR2	أعطى, to give
- Transitive 3	V+ CMP +TR3	أرى, خَبَّر, حدّث
- Intransitive, لازم	V+ CMP +ITRS	مات, قام, dead, maata
- قاصر	V+ CMP +ITRO	سقط, to fell
- Incomplete Verb, الفعل الناقص	V+INC	كان, was, kaana
- Approximation Verb, أفعال المقاربة	V+INC +APR	كاد ، أوشك ، كرب
- Trust Verb, أفعال الرجاء	V+INC +TRU	حرى, اخلولق, عسى
- Begging Verb, أفعال الشروع	V+INC +BEG	بدأ, طفق
- كان و أخواتها	V+INC +KNA	كان , ظل, ليس
- Particle, حرف	PART	
- ملحقة بالإسم	PART+n	
- Preposition, حروف الجر(حروف الإضافة)	PART+n+ADD	على, في
- etc.		
- ملحقة بالفعل	PART+v	
- Futurity particles, حروف الإستقبال	PART+v+FTR	س, سوف, will
- Condition particles, حروف الشرط	PART+v+CDT	لولا, لو, إنْ, if
- etc.		
- ملحقة بالإسم و بالفعل	PART+b	
- Conjunctions , حروف العطف	PART+b+CNJ	و, ثم, بل

3   Simple Arabic Verbal Sentence Parsing

To parse an Arabic sentence, we must study its grammatical structure. In the first sub-section, we present the general structure of the simple Arabic sentence, and in the second sub-section, we study the simple Arabic verbal sentence's grammatical structure. The third sub-section is devoted to morphological disambiguation. Additionally, the last sub-section deals with the parser implementation in NooJ platform.

3.1   Simple Arabic Sentence Study

The simple sentence, in every language, is formed from two required elements; these two elements are the predicate (المسند, al-musnad) and the subject (المسند إليه, al-musnad 'ilayh) [16–18, 23–25]. In the Arabic language, the subject can be dropped (pro-drop). The relationship holding between the predicate and the subject is called the attribution (al-'isnād, الإسناد). The simple Arabic sentence includes one event (ḥadaṯ wāḥid, حدثواحد) or a singular attribution. The predicate and the subject constitute a predicative kernel that builds the sentence meaning. In the Arabic language, the predicate can be a complete verb, in this case, the sentence is verbal, or a noun, in this case, the sentence is nominal, but the subject is always a noun phrase (مركب إسمي). The predicate can be governed by some particles affecting the verb, such as negation particles, or the noun such as annulling particles. The governor particle is called the head of the sentence. Incomplete verbs are regarded as head both in the verbal and the nominal sentence. Direct object(s) and indirect object(s) are regarded as complement, (الفضلة, al- faḍlah). This component is optional such as the head. Arabic grammarians have established the following formula, describing the general structure of the simple sentence:

$$الجملة = [\ الصدر \ ] \ (المسند و المسند إليه) \ [\ الفضلة \ ] \qquad (1)$$

The sentence, al-ğumlah = [the head, al-ṣṣadr] *(the predicate, al-musnad and, wa the subject, al-musnad 'ilayh)* [the complement, al-faḍlah]

The simple Arabic verbal sentence can not exceed four main components. The first one, the head, can be replaced by some particle sub-classes such as negation or interrogation particles, or by incomplete verbs. The sentence kernel includes the predicate and the subject, select zero, one or many accusative forms (complement) to achieve the sentence meaning. These four components have a free order in the simple Arabic sentence (both verbal and nominal). Figure 1 shows an example of the negative simple Arabic verbal sentence:

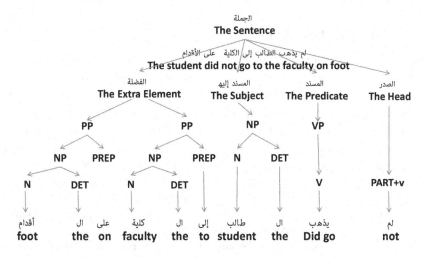

Fig. 1. An example of simple verbal sentence structure

3.2 Grammatical Structure of Simple Arabic Verbal Sentence

The simple Arabic verbal sentence contains one predicative kernel, and its predicate is a complete verb whatever its position is in the sentence. In the Arabic language, sentence components order is free; we can match various combinations: SPC (Subject-Predicate-Complement) (الطالب فهم الدرس, the student understood the lesson), PSC (فهم الطالب الدرس), PCS (فهم الدرس الطالب), etc.

Regarding transitivity feature, Arabic verbs are classified into five main subclasses (see Sect. 2.1). All verbs of a given subclass share the same syntactic structure, and then we must match five different syntactic structures of the simple Arabic verbal sentence: the first one is based on (قاصر) verbs; in this structure the predicate and its subject (object in the meaning) are only required in the structure, such as in (سقطت التفاحة. The apple fell, and انكسر الزجاج, the glass was broken). The second structure is based on intransitive verbs; only the predicate and the subject are mandatory in the structure, such as in (نام الطفل, the child slept). The third one is constituted by transitive verbs to one object; then we must have three required syntactic positions in this structure: predicate, subject and a required complement such in (يشرح الأستاذ الدرس, the teacher explains the lesson and يشير المنبه إلى الواحدة, it is one o'clock). The forth structure is based on transitive verbs to two objects; in addition to the predicate and the subject, a complement including two objects is required, such as in (أعطى علي أخيه رسالة, Ali gave a letter to his brother and وهب الملك الوزير وساما, the king granted a medal to the minister). The last structure of the simple Arabic verbal sentence is obtained using transitive verbs to three objects; the complement must include three accusative forms, such as in (أخبر العميد الجنود العدو قادما, the brigadier reported to the soldiers that the enemy is coming).

An optional part of the complement, which includes one or more prepositional/locative phrases, can be added to required components, in the five syntactic structures of the simple Arabic verbal sentence, and it gives a particular meaning to the sentence such as in (نام الطفل على السرير في البيت, the child slept on the bed at home, يشرح الأستاذ الدرس في القسم, the professor explains the lesson in the class).

3.3 Morphological Disambiguation

Arabic language is ambiguous at the morphological level; the same undicritized word can have many meanings. In the sentence "لقد نفع علم الأستاذ التلاميذ في حل المسألة", the word نفع without dicritization can be the verb نَفَع (to benefit), the verb نَفّع (to make some body benefit), the nouns نَفْع, نِفَع, the verbal noun نَفْع (benefit). The word علم can be the noun عِلْم (science), the noun عَلَم (banner), the verb عَلِمَ, (to know), the verb عَلَّمَ (to learn) or the verb عَلَمَ (to notify). In this sentence, we have another ambiguous word; the word حل can be the noun حَل (solution), or the verb حَلّ (to resolve). Ambiguity in the first word can be solved using two constraints: the first one is "after a particle affecting on the verb, we must have a verb", thus after قد (indeed), that affects the verb, we must have a verb. The second constraint is "we cannot have two consecutive complete verbs", and then حل is a noun because before it we have a complete verb. Ambiguity, in the last ambiguous word حل, can be solved using the constraint "after a particle affecting on noun, we must have a noun". Thus, في (in) is a preposition affecting only on noun; then the word حل must be a noun. Arabic morphological ambiguity can be solved using Arabic lexical and syntactic rules.

Our disambiguation method is applied after the morphological analysis and before the syntactic analysis. The morphological analyzer produces all possible interpretations of the textual Arabic word. Disambiguation would be resolved by applying certain types of lexical and syntactic constraints that are defined with local grammar rules [9–12, 16] (see Figs. 2, 3 and 4). These rules lead to a correct parse. It could resolve morphological ambiguity.

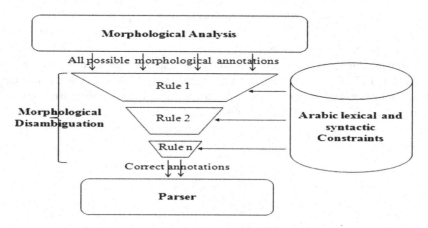

Fig. 2. Morphological disambiguation schema

After that, we implement a set of disambiguation rules. These rules model Arabic lexical and syntactic constraints. They are implemented as local grammars using the NooJ platform. Each analyzed sentence is matched with these grammars in a sequential mode in order to overcome useless morphological annotations. The following local grammars (Figs. 4 and 5) summarize disambiguation rules.

Fig. 3. Disambiguation rule 1

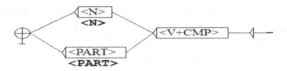

Fig. 4. Disambiguation rule 2

3.4   Parser Implementation

Arabic is an agglutinative language; its syntactic parsing requires the morphological analysis at the first step. In a previous work [15], we have already implemented a set of morpho-syntactic graphs processing agglutination in NooJ platform.

Arabic verbs can produce five different grammatical structures according to the transitivity feature, presented above. We have implemented five sub-grammars covering simple Arabic verbal sentence grammar. Arabic particles affecting the verb are classified into 12 sub-classes; then we have implemented (12 + 1) * 5 + 5 = 70 simple Arabic verbal sentence types, all possible structures and types (affirmative, negative, interrogative, etc.) of the simple Arabic verbal sentence in the active voice. Our grammar consists of forty syntactic graphs implemented in NooJ platform. The following figure describes the first level of our grammar.

Fig. 5. First level of our grammar

The first level of every sub-grammar handles and annotates the sentence main components. We reduce the syntactic ambiguity just after the generation of possible structures of the input sentence, using verb-subject agreement constraints and some semantic features such as rationality. In the rest of this section, we present the sub grammar based on the transitive verb to one object of the simple Arabic verbal sentence. Figure 6 presents the sub-grammar based on the transitive verb to one object of the simple Arabic verbal sentence:

The predicate of this kind of sentence is a transitive verb to one object (V + CMP + TR1). The head, in the verbal sentence, is a governing particle in the verb or an incomplete verb (فعل ناقص). Figure 7 shows the syntactic sub-classes replacing the head in the Arabic verbal sentence.

The subject is always a noun phrase (مركب إسمي). The complement, in this type of Arabic sentence, must include a noun phrase (Direct Object Complement), or at least a

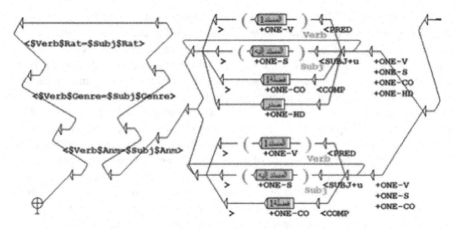

Fig. 6. Sub grammar based on transitive verb to one object

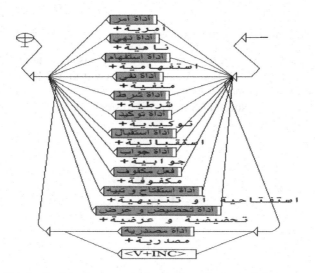

Fig. 7. Particle classes replacing the head in the verbal sentence

prepositional/locative phrase (Indirect Object Complement). The complement can be extended by one or many prepositional/locative phrase(s) such as in:

كتب الطالب الدرس [بالقلم على الورق], the student wrote the lesson [with a pen on paper]. In this case, the complement gives a particular meaning to the sentence.

4   Parser Tests

To test the syntactic parser in Arabic corpora, we must segment corpora text in sentences. This module requires a specific study that is not the aim of this study. Then, we have created manually a text of one thousand simple verbal sentences representing all possible structures of this type of the Arabic sentence. Our morphological disambiguation grammar and syntactic parser are then applied to analyze a selected text. The number of totally disambiguated sentences is six hundred and seventy (67%). However, we still have ambiguity in some sentences due to the ambiguity feature of Arabic language. The number of partially disambiguated sentences is one hundred and fifty cases (15%). Then, the rate of disambiguation is around eighty-two percent. This can be explained because some constraints are not yet implemented. This issue could be solved once we implement a semantic analyzer beside the syntactic analyzer. Regarding the parsing task, the number of successfully parsed simple verbal sentences is nine hundred and twenty. The rate of our analyzer is around ninety-two percent. This is obvious since some grammars are not yet implemented. Figure 8 presents the tagging of a sentence just before the disambiguation step. Figure 9 presents the tagging of the same sentence just after the disambiguation step and before the parsing step. Figures 10 and 11 contain the syntactic analysis result for the same sentence. We notice the success of the analysis even though the sentence components' order is not the same in each of them.

4.1   Morphological Disambiguation Test

In the example cited in Sect. 4, the sentence is ambiguous at the morphological level. We show in Fig. 8 the NooJ Text Annotation Structure (TAS) before applying our disambiguation local grammar on this sentence.

Figure 9 shows the NooJ TAS after the disambiguation step; all impossible annotations are filtered out.

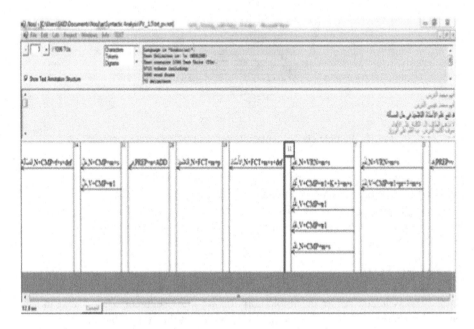

Fig. 8. NooJ TAS before disambiguation

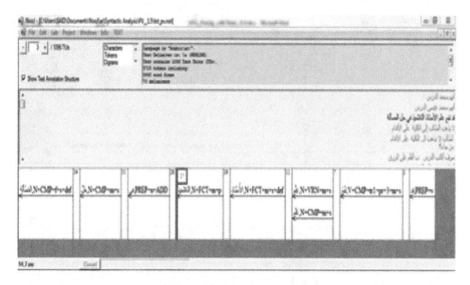

Fig. 9. NooJ TAS after disambiguation

4.2 Parsing Test

After the morphological disambiguation step, we obtain a disambiguated sentence which is the input of our parser. The parser has to match the parse tree(s) to the input sentence. Figures 10 and 11 show the NooJ TAS after the parsing step applied to two sentences. These sentences are similar but the order of their components is different.

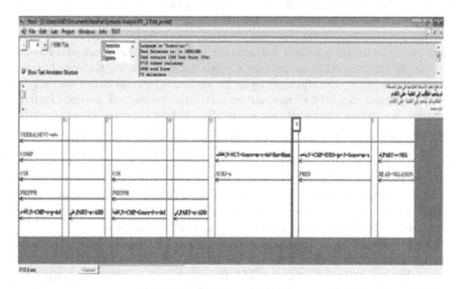

Fig. 10. NooJ TAS after parsing

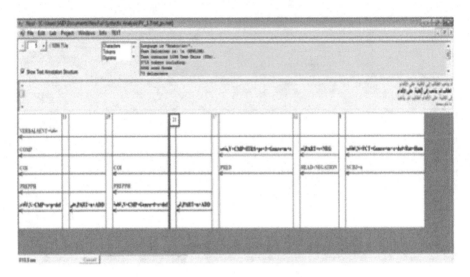

Fig. 11. NooJ TAS after parsing

The parser returns the same syntactic annotations of the sentence components in the two proposed cases.

5 Related Work

In the previous literature, many approaches were applied to implement a syntactic analyzer for parsing Arabic sentences. Actually, there are three main approaches: linguistic, statistical, and hybrid. The linguistic methods are based on lexicon and grammar rules such as in [1–3]. This approach lacks resources (dictionary, grammar, etc.). For instance, the Arabic grammars do not cover all sentences' types. The works based on the statistical approach use annotated corpora such as Treebank (TB) and approximate grammatical rules from the corpora parse trees using automated quantitative methods [4, 5]. The shortcoming of the statistical methods is that they rely on reference corpora. So, if the reference corpora do not cover all possible sentence structures, we cannot obtain reliable results. The hybrid approach incorporates linguistic rules and corpora-based statistics. We can cite in this approach [6, 7].

In [1], Ouersighni et al. implemented a parser for unrestricted Arabic sentences using the AGFL (Affix Grammars over Finite Lattice) system. AGFL grammars are a restricted form of Context Free grammars (CFG). Context-Free production rules are extended with features (affixes) for implementing agreement in the sentence. These features are passed as parameters to the grammar rules. Disambiguation is not resolved in this paper. This parser parses both nominal and verbal sentences.

In [2], Abuawad et al. developed an Arabic parser based on analyzing the Arabic language grammar conforming to gender and number. The author formalized grammar rules using Context Free Grammar (CFG) formalism, and implemented a top–down algorithm parsing technique with recursive transition network.

In [3], Al-Taani et al. presented a top-down chart parser for parsing simple Arabic sentences using Context Free Grammar (CFG) formalism. The authors tested the proposed parser on 70 sentences extracted from Arabic real-world documents.

In [4], Shahrour et al. presented a methodology of using models with access to additional information of exact syntactic analysis and rules to offer an enhanced estimation of case and state. The expected case and state values are, then, used to re-tag the Arabic morphological tagger MADAMIRA output by choosing the best match to its graded morphological analysis. Since what they are learning to expect is how to correct MADAMIRA's baseline choice (as opposite to a generative model of case-state). They also re-applied the model on its output to repair mainly spread agreement errors.

In [5], Khoufi et al. suggested an approach for parsing Arabic sentences based on supervised machine learning using Support Vector Machine (SVMs). This system selects syntactic labels of the sentence. This proposed method has two steps: the first one is the learning step and the second is the prediction step. The first step is based on a training corpus, extraction features, and a set of rules that are obtained from the corpus of learning. The second step implements the results of learning obtained from the first stage to accomplish parsing.

In [6], Ibrahim M. et al. proposed a hybrid system composed of the statistical and linguistic approach for Arabic grammar analysis, parsing and resolving end word cases. This system showed an adequate accuracy and it is easy to be implemented. However, the system requires deep knowledge of Arabic despite the use of learning portions availability. The authors use a set that contains 600 Arabic sentences to make system experiments.

In [7], Khoufi et al. implemented the Arabic sentence parser based on Probabilistic Context Free Grammar (PCFG). The authors proposed a method that consists of two phases: in the first one a PCFG is induced from Arabic Treebank parse trees, and in the second, the authors implemented the Viterbi parsing algorithm using the induced grammar in the first phase. The authors have tested the parser on 1650 sentences extracted from the same Treebank.

Arabic text linguistic analysis can concern different analysis levels (morphological, syntactic, semantic and pragmatic); every level uses the previous one's result (output). The treatment is applied sequentially in the text (transduction on text automata), in order to resolve ambiguity. Then, the result must be unified between the linguistic analysis levels to have easy communication between them. NooJ guarantees high integration of all levels of natural languages' description thanks to compatible notations and a unified representation for all linguistic analysis results, which enable different analyzers at different linguistic levels to communicate with one another. [9, 11]. By using this platform, we can implement a set of analysis tools applied in cascade in the input text.

6 Conclusion and Perspectives

In this paper, we have presented our methodology of the simple Arabic verbal sentence parsing. This methodology consists of classifying the entries of an Arabic electronic dictionary, regarding syntactic and semantic classes, implementing morphological disambiguation rules, and creating syntactic grammars. The implementation is performed using the NooJ platform. If the platform NooJ allows to process all the stages of analysis (morphological, syntactic and semantic). Our work will focus on the stage of syntactic analysis, with a preliminary stage of disambiguation.

Thus, we have implemented many transducers (morphological disambiguation rules) modeling a set of lexical and syntactic constraints in Arabic language. These transducers are applied sequentially. After that, with our structural grammars, we have analyzed several simple Arabic verbal sentences, disambiguated them automatically and generate their annotated parse tree(s).

Our method will not be limited to the simple verbal sentence; we will extend it to other types of the Arabic sentence. Therefore, we will be able to syntactically analyze different texts and corpora thereafter.

Once the Arabic analyzer is done, many issues could be solved such as automatic diacritics, Arabic sentences' correction, and accurate translation. Also, other disambiguation rules could be implemented when the semantic analysis can be used.

References

1. Ouersighni, R.: Robust rule-based approach in Arabic processing. Int. J. Comput. Appl. **93** (12), 0975–8887 (2014)
2. Abu Awad, A., Hanandeh, E.: Developing a transition parser for the Arabic language. Int. J. Adv. Comput. Sci. Appl. (IJACSA) **7**(9), 173–175 (2016)
3. Al-Taani, A., Msallam, M., Wedian, S.: A top-down chart parser for analyzing Arabic sentences. Int. Arab J. Inf. Technol. **9**(2), 109–116 (2012)
4. Shahrour, A., Khalifa, S., Habash, N.: Improving Arabic diacritization through syntactic analysis. In: Proceedings of the Conference on Empirical Methods in Natural Language Processing, Lisbon, Portugal, pp. 1309–1315, September 2015
5. Khoufi, N., Louati, S., Aloulou, C., Belguith, L.H.: Supervised learning model for parsing Arabic language. In: Proceedings of the 10th International Workshop on Natural Language Processing and Cognitive Science (NLPCS), Marseille, France, pp. 129–136 (2013)
6. Ibrahim, M., Mahmoud, M., El-Reedy, D.: Bel-Arabi: advanced Arabic grammar analyzer. Int. J. Soc. Sci. Hum. **6**(5), 341–346 (2016)
7. Khoufi, N., Aloulou, C., Belguith, L.H.: Parsing Arabic using induced probabilistic context free grammar. Int. J. Speech Technol. **19**(2), 313–323 (2016). Springer US
8. Blanchete, I., Mourchid, M., Mouloudi, A., Mbarki, S.: Formalizing Arabic inflectional and derivational verbs based on root and pattern approach using NooJ platform. In: Proceedings of the International NooJ Conference, NooJ 2017, Kenitra-Rabat, Morocco (2017)
9. Silberztein, M.: Formalizing Natural Languages The NooJ Approach. ISTE Editions, London (2016)
10. Silberztein, M.: Syntactic parsing with NooJ. In: Proceedings of the International NooJ Conference, Tozeur, Tunisia (2009)
11. Silberztein, M.: NooJ Manual (2003). www.nooj4nlp.net
12. Silberztein, M.: Disambiguation tools for NooJ. In: The International NooJ Conference, Budapest, Hungary (2008)
13. Mourchid, M., EL Faddouli, N., Amali, S.: Development of lexicons generation tools for Arabic: case of an open source conjugator. Int. J. Nat. Lang. Comput. (IJNLC) **5**(2), 13–25 (2016)

14. Mourchid, M.: Génération morphologique et applications. Thèse de Spécialité de 3ème Cycle, Université Mohammed V, Juillet 1999
15. Kassmi, R., Mourchid, M., Mouloudi, A., Mbarki, S.: Processing agglutination with a morpho-syntactic graph using NooJ. In: Proceedings of the International NooJ 2017 Conference, Kenitra-Rabat, Morocco, May 2017
16. Bourahma, S., Mbarki, S, Mourchid, M, Mouloudi, A.: Disambiguation and annotation of Arabic simple nominal sentences using NooJ platform. In: Proceedings of the International NooJ 2017 Conference, Kenitra-Rabat, Morocco, May 2017
17. Assamirai, S.F.: Composition and Types of Arabic Sentence, 2nd edn. dar al kitab, Bagdad (2007)
18. Alsuhaibani, S.O.: The verbal sentence in written Arabic. Thesis for the degree of Doctor of Philosophy, University of Exeter, Ukraine, April 2012
19. Gibson, E.: Linguistic complexity: locality of syntactic dependencies. Cognition **68**, 1–76 (1998). Elsevier
20. ابن الناظم : شر ح ابن عقيل على ألفية ابن مالك,الجزء الأول, الجزء الثاني, دار التراث, القاهرة, مصر (1980)
21. أبو زيد المقرئ الإدريسي : حروف المعاني في اللغة العربية دراسة تركيبية ودلالية, مؤسسة الإدريسي ,الدار البيضاء (2016)
22. الحسين بن قاسم المرادي: الجنى الداني في حروف المعاني, دار الآفاق الجديدة, بيروت (1983)
23. الرازي فخر الدين : نهاية الإيجاز في دراية الإعجاز, مطبعة الآداب ، القاهرة (1317 ه)
24. سيبويه: الكتاب, بولاق, القاهرة, (1316 ه)
25. ابن كمال الباشا: أسرار النحو. الطبعة الثانية ,دار الفكر للطباعة و النشر و التوزيع, فلسطين (2002)

Identification of Croatian Light Verb Constructions with NooJ

Krešimir Šojat[1], Kristina Kocijan[2(✉)], and Božo Bekavac[1]

[1] Department of Linguistics, Faculty of Humanities and Social Sciences,
University of Zagreb, Zagreb, Croatia
`{ksojat,bbekavac}@ffzg.hr`
[2] Department of Information and Communication Sciences,
Faculty of Humanities and Social Sciences,
University of Zagreb, Zagreb, Croatia
`krkocijan@ffzg.hr`

Abstract. This paper deals with the identification of verbal multiword expressions in Croatian using NooJ. We focus on light verb constructions, i.e. constructions consisting of a light verb and a nominal part (e.g. *donijeti odluku* 'to make a decision' or 'to reach a decision'). The nominal part in light verb constructions (LVCs) can be either a noun phrase (e.g. *dati udarac* 'to give a blow') or a prepositional phrase (*staviti na raspolaganje* 'to put at disposal'). Light verbs are entirely or partially deprived of their lexical meaning and the meaning of the whole construction is conveyed by their complimentary NPs or PPs. All elements in these constructions act as a single unit at some level of linguistic analysis, particularly at the level of semantics and, more recently, of syntax.

Keywords: Verbal multiword expressions · Light verb constructions
Syntactic grammar · Croatian · NooJ

1 Introduction

This paper deals with the identification of verbal multiword expressions in Croatian using NooJ. We focus on light verb constructions (LVCs), i.e. constructions consisting of a light verb and a nominal part (*donijeti odluku* 'to make a decision' or 'to reach a decision'). The nominal part in light verbs constructions is usually either a noun phrase (e.g. *dati udarac* 'to give a blow') or a prepositional phrase (*staviti na raspolaganje* 'to put at disposal'). Light verbs are entirely or partially deprived of their lexical meaning and the meaning of the whole construction is conveyed by nominal parts. All elements in these constructions act as a single unit at some level of linguistic analysis, particularly at the level of semantics and syntax.

Typically, LVCs can be fixed or flexible. LVCs are fixed in terms of the fact that the paradigmatic choice of elements and their syntactic arrangement cannot be altered. Croatian LVCs are flexible to a certain extent. Syntactically, light verbs are inflected and they can be passivized. They are also marked as perfectives or imperfectives. In some LVCs, nouns can be used in singular or plural forms and/or in different cases. An

© Springer International Publishing AG 2018
S. Mbarki et al. (Eds.): NooJ 2017, CCIS 811, pp. 96–107, 2018.
https://doi.org/10.1007/978-3-319-73420-0_8

important feature of LVCs is that they can frequently be substituted with a single "heavy" verb (*donijeti odluku* 'to reach a decision' –> *odlučiti* 'to decide').

The identification of verbal multiword expressions is an important task in numerous NLP tasks [3]. However, their identification and annotation have so far gained little attention in Croatian corpora. [11] report on the annotation of various types of verbal multiword expressions (MWEs) in Croatian treebanks. The result of this effort is the database of verbal MWEs annotated and extracted from a dependency treebank for Croatian [1]. The database is structured in a way that queries are possible over whole MWEs as well as their individual components.

The objectives of this paper are to develop NooJ grammars for the recognition and extraction of LVCs in morphosyntactically tagged Croatian corpora, to further develop and enrich the existing database of MWEs. In this research, we start from the nominal parts of light verb constructions, i.e. NPs and PPs. These phrases can appear in various LVCs, although not always in the same form. For example, the noun *raspolaganje* 'disposal' in PPs occurs in singular/accusative case in the LVC *staviti na raspolaganje* 'to put at disposal' and in singular/locative case in the LVC *biti na raspolaganju* 'be at disposal'. Grammatical cases are influenced by the choice of a particular verb. Starting from the ten most frequent NPs and PPs from the LVCs recorded in the aforementioned database, we want to determine the full range of light verbs that co-occur with particular NPs or PPs from the database[1]. Another task is to determine the full range of morphosyntactic variants of selected NPs and PPs.

In order to enable this process, we apply previously developed NooJ resources for Croatian, namely dictionaries and morphological grammars [12, 13] and a set of local grammars for the morphosyntactic and syntactic analysis of Croatian built explicitly for this project.

Obtained results are valuable for the further enrichment of the database of Croatian verbal MWEs, for the development and refinement of existing NooJ dictionaries and grammars for Croatian, as well as for a more precise definition and delimitation of this concept in Croatian linguistic theory.

The paper is structured as follows: in Sect. 2, we present the major types of multi-word expressions (MWEs) and criteria for their classification. Section 3 deals with verbal MWEs and focuses on light verb constructions. In Sect. 4, the annotation of LVCs in a treebank is presented. Sections 5 and 6 describe the design of NooJ dictionary and a syntactic grammar for the recognition of LVCs. Section 7 brings the result of the applied method. The final part of the paper provides a conclusion and an outline for future works.

2 Multiword Expressions

Light verb constructions are conceived and described in various approaches as a part of a broad group of various multiword expressions (MWEs). One of the main features of such expressions is that they cannot be interpreted semantically, syntactically or

---

[1] However, although this is one of our final objectives, we will deal with this part in future work.

perhaps at some other level, unless they are treated as a whole. In other words, multiword expressions refer to various types of constructions consisting of two or more words that act as a single unit at some level of analysis. [8] defines such expressions as "idiosyncratic interpretations that cross word boundaries (or spaces)". The authors give an extensive account of various MWEs in English and list criteria for their classification. MWEs are divided into those that are fixed, i.e. the selection of the elements that can occur in these MWEs and their syntactic order are never altered, and those that can be modified to a certain degree. The modification can encompass either morphosyntactic properties of elements and/or their selection. Semantically, MWEs can vary from more or less compositional to completely idiosyncratic. MWEs are usually divided into noun compounds, multiword named entities, different types of complex verb phrases, idioms and others.

Reporting on various criteria for the classification and annotation of MWEs in corpora and treebanks, [2, 7] divide MWEs into the following groups: 1. nominal MWEs, 2. verbal MWEs, 3. prepositional MWEs, 4. adjectival MWEs, 5. MWEs of other categories, 6. proverbs.

The annotation and recognition of MWEs in Croatian corpora have so far gained little attention, although these constructions are a challenge in various NLP tasks. In this paper, we focus on verbal MWEs in Croatian. We deal with this type of MWEs because (a) there is little research done on the identification and annotation of verbal MWEs in Croatian language resources so far and (b) there is no public resource that would enable a comprehensive research of Croatian MWEs. The only corpus annotated for verbal MWEs is described in [11]. We shall describe this resource in more detail in Sect. 4. Before that, in Sect. 3 we present a brief description as well as some criteria for the classification of verbal MWEs in Croatian.

3   Verbal MWEs

In [2] as well as [7] verbal MWEs are divided into the following types:

(1) phrasal verbs,
(2) light verb constructions,
(3) VP idioms and
(4) other verbal MWEs.

We will briefly discuss groups 1–3 and describe their status and description in Croatian literature in this area.

First, the category of phrasal verbs is generally neither recognized nor discussed in Croatian grammars and reference books. In recent research, [5] stress that particular prepositions that appear within VPs can significantly change the meaning of a verb. It is argued that such expressions should therefore be treated as a single unit since the meaning of a verb that co-occurs with an object PP can significantly differ from the meaning of the same verb that co-occurs with an adverbial PP, for instance: 1. *otići* 'to go' vs. 2. *otići po* 'to fetch'.

Second, light verb constructions (e.g. *donijeti odluku* 'to make a decision, to reach a decision', *dati poljubac* 'to give a kiss') consist of a verbal and a nominal component. The nominal component contains an NP or a PP. If it is an NP, it is usually in the accusative case. In terms of word-formation, nouns in NPs are frequently derived from verbal stems (e.g. *stajati na raspolaganju* 'be at disposal'; the noun *raspolaganje* is derived from the verb *raspolagati* 'to dispose'). In terms of semantics, light verbs are entirely or partially deprived of their lexical meaning. In other words, NPs or PPs actually express the meaning of the whole construction. Light verb constructions (LVCs) in Croatian are syntactically flexible to a certain degree. Light verbs can be inflected, passivized and marked as perfectives or imperfectives. In some LVCs, nouns are used both in singular and plural forms and/or in different cases. As in many languages, Croatian LVCs can frequently be substituted with a single "heavy" verb, e.g. *donijeti odluku – odlučiti* 'to make a decision' – 'to decide', although this is not possible in many cases [4].

Third, VP idioms (or *phrasemes*, as they are usually called in Croatian reference books, [6]) are usually subdivided into two groups: decomposable and non-decomposable idioms. The division is based on the degree of semantic and syntactic opaqueness of the whole construction in regard to its elements, as well as on the possibility of the word order change within an idiom [3].

All verbal MWEs listed above form complex sentence predicates, i.e. multiword units, and therefore need to be identified and annotated in Croatian language resources. As it has been mentioned above, [11] reported on the annotation of verbal MWEs in a Croatian corpus. This work was done on the Universal Dependency (UD) Treebank for Croatian. In Sect. 4, we will briefly present this language resource and the applied annotation scheme.

4 Procedure

There are three dependency treebanks available for Croatian:

(1) Croatian Dependency Treebank (HOBS) with 4,626 sentences of Croatian newspaper. HOBS is freely available for on-line search (hobs.ffzg.hr).
(2) SETIMES.HR dependency treebank (http://nlp.ffzg.hr/resources/corpora/setimes-hr/) built on top of the newspaper text from the SETIMES parallel corpus. The treebank contains approximately 9,000 sentences. Both treebanks are annotated with modified versions of schemes used in the Prague Dependency Treebank project.
(3) Universal Dependency (UD) Treebank for Croatian [1]. This treebank was also built from SETIMES parallel corpus, but annotated according to UD annotation. The version of the UD treebank that was used consisted of 3,557 sentences.

The UD Treebank was chosen for our work. Prior to annotation in the treebank, a list of verbal MWEs from available work done in this area for Croatian was compiled. An initial list of 80 LVCs was taken from [10] as well as [4], and used to determine the criteria for annotation in the treebank.

In the second step, 3,557 sentences from the UD treebank were manually annotated for Croatian phrasal verbs, LVCs, VP idioms and other verbal MWEs (e.g. multiword predicates). Verbal MWEs were marked in the corpus on a separate level of annotation. A detailed account of the whole procedure and results are presented in [11]. Further, we will discuss only the results concerning LVCs.

The total number of annotated LVCs in the treebank was 847. In the next step, LVCs were extracted from the treebank. The obtained results were structured into a database which enables queries over particular light verbs or nominal parts (NPs and PPs). Thus, we were able to determine which light verbs are predominantly used, which nominal parts they combine with, the frequency of particular NPs and PPs, etc. The database structured in this way also enabled the development of NooJ grammars for the recognition of LVCs in unannotated texts. We will deal with this issue in the next section.

5 LVC Dictionary Design

Prior to constructing any syntactic grammars in NooJ [9], that will allow us to locate and annotate LVCs, their detailed analysis had to be performed in order to build a dictionary that will be supportive of our grammars. LVCs were analyzed for the type of the phrase and the number of elements they contain. In other words, LVCs were divided into different categories depending on whether they consist of a light verb and an NP, a light verb and a PP, a light verb and two PPs, etc. The analysis yielded five different types of LVCs (Fig. 1). We list here the types we established:

1. verbs with the prepositional phrase – *dobiti na težini* ('to gain on weight')
2. verbs with the nominal phrase and a preposition – *baciti sumnju na* ('to cast doubt on')
3. verbs with the prepositional phrase and a preposition – *biti u skladu s* ('to be in accordance with')
4. verbs with a nominal and prepositional phrases – *izdati nalog za uhićenje* ('to issue a warrant for arrest')
5. verbs with the nominal phrase – *doživjeti rast* ('to achieve growth').

The verb of an LVC was chosen as a main entry in the dictionary while complements were entered as its attribute values. Complements refer to various combinations of nominal elements. Five different attributes were introduced to describe the verb's complements:

- PREP is used to introduce the preposition
- PREPX is used to introduce the NP used inside the PP
- SUFX is used to introduce the standalone NP
- SUFXB is used to introduce the noun phrase of a PP where both the NP and PP are present
- AFIX is used to introduce the preposition of a PP where an additional preposition exists (Type 3).

```
#########
# LVCType1 with PP
##################
dobiti,V+FXC+LVCType1+PREP=na+PREPX=težina+FLX=ČUTI

#########
# LVCType2 with NP + preposition
###############################
baciti,V+FXC+LVCType2+SUFX=sumnja+PREP=na+FLX=BACITI

#########
# LVCType3 with PP + preposition
###############################
poslužiti,V+FXC+LVCType3+AFIX=kao+SUFX=temelj+PREP=za+FLX=UGOJITI

#########
# LVCType4 with NP  + PP
########################
izdati,V+FXC+LVCType4+SUFX=nalog+PREP=za+SUFXB=uhićenje+FLX=IZDATI

#########
# LVCType5 with NP
#################
bilježiti,V+FXC+LVCType5+SUFX=rezultat+FLX=BILJEŽITI
```

Fig. 1. Model of an LVC dictionary and its five categories with examples

Since the main entry of an LVC is a verb, it also carries the marker V as a word type category. In order to differentiate it from regular verbs, additional NooJ's special feature denoting frozen expression construction, +FXC, is introduced as well. However, to further differentiate among different LVCs, there is an additional marker attached, denoting its LVC type (LVCtype1–LVCtype5). This marker is used within the syntactic grammar (cf. Sect. 6) to recognize entire LVC constructions.

The number of dictionary entries, according to their LVC type, is given in Table 1, together with the actual number of light verb constructions that we are able to recognize.

Table 1. The distribution of dictionary entries vs LVCs.

LVC type	# of entries	# of LVCs
1	53	91
2	45	51
3	5	5
4	4	4
5	191	315
Total	298	466

Notice that these numbers differ, especially for the types 1 and 5. This is because the same verb may enter different expressions, i.e. the verb is the same, but the noun (in either the standalone NP or the one that is part of a PP) or a preposition can be different.

The verb *voditi* ('to lead') is an example that explains this. It can occur in four different LV constructions each having a different meaning.

```
voditi,V+FXC+LVCType5+FLX=SUDITI        :        to lead
     +SUFX=borba                               +SUFX=fight
     +SUFX=razgovor                            +SUFX=conversation
     +SUFX=pregovor                            +SUFX=negotiation
     +SUFX=diskusija                           +SUFX=discussion
```

So far, the verb *imati* ('to have') is the most productive Type 5 light verb, with 32 possible expressions.

Similar logic has been followed to describe different expressions in Type 1 constructions. However, since there are two possible attributes that can change (a preposition and a noun that follows), a new entry is made for the same verb but with a different prepositional attribute value.

```
(1) dovesti,V+FXC+LVCType1+FLX=POJESTI   :        to bring
        +PREP=do                              +PREP=to
          +PREPX=jačanje                        +PREPX=strength
          +PREPX=sukob                          +PREPX=conflict
          +PREPX=zbacivanje                     +PREPX=overturn

(2) dovesti,V+FXC+LVCType1+FLX=POJESTI   :        to bring
        +PREP=u                               +PREP=into
          +PREPX=opasnost                       +PREPX=danger
          +PREPX=pitanje                        +PREPX=question
```

Nouns in all LVC types are entered in the dictionary as a singular nominative noun whereas their actual number and case, as well as possible inversions and insertions, are recognized via syntactic grammars. In the next section, we will describe the syntactic grammar used for the recognition and annotation of LVCs in text and show how we used dictionary codes inside the grammar.

6　LVC Syntactic Grammar

Only one syntactic grammar was designed to recognize and annotate Croatian LVCs. It uses the logic adopted in the dictionary. Thus, the main grammar has five subgraphs each describing one LVC type category. There are also 3 additional subgraphs that are reused among each type. They are dealt with in (Fig. 2):

- auxiliary verbs (auxiliary verbs *to be* and *to have* for complex VPs, negation and the reflexive pronoun 'se') in <PG> subgraph,
- word categories that may appear before a noun (adjectives, pronouns, numbers) in <beforeN> or
- word categories that may appear after a noun (1 or more NPs in genitive) in <afterN>.

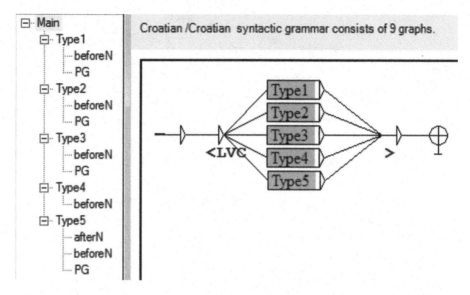

Fig. 2. The main grammar with its structure

We will proceed with a detailed description of subgraphs for Type 1 and Type 5 LV constructions.

The recognition of Type 1 LVCs (Fig. 3) can take two directions, depending on whether or not the main verb is in the first position (upper path) or in the last one, i.e. in the inversion (lower path). If we take an upper path, the light verb <V+LVCType1> is followed by a preposition recognized via VPREP (i.e., if variable $S exists as an attribute $PREP for a variable $V in the dictionary). After the preposition, the second attribute value (marked in the dictionary as PREPX) is recognized as either a noun or a pronoun. These elements should morphosyntactically agree with the case required by the preposition[2], recognized via variable $S. The node <$S$Case=$N$Case> checks whether this condition is satisfied.

If the verb is in inversion, i.e. preposition (stored in variable $S) and a noun/pronoun (stored in variable $N) preceed it, we take the lower path of the graph. In this case, after we recognize a preposition, noun/pronoun and a light verb (always in this order of appearance), we must verify that the preposition and a noun/pronoun match in case, but also that such a preposition and a noun/pronoun exist as attributes PREP <$S=$V$PREP> and PREPX <$N_N+Nom+s=VPREPX> for the recognized verb. If any of these verifications fails, the string will not be marked as an LVC.

Regardless of the path we take, an auxiliary verb (described in:PG subgraph) may appear before the main verb while the noun may be preceded by a number of adjectives, a pronoun and a number (described in:beforeN subgraph).

---

[2] It is important to notice that the preposition in Croatian language do not have a case. However, each one only allows NPs in some cases to follow them. This information is incoded in the dictionary of pronouns as a Case attribute and it is what we refer to at this place.

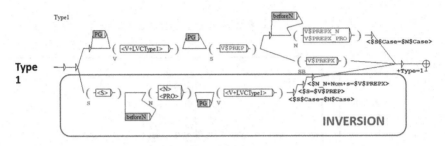

Fig. 3. Subgraph for the recognition and annotation of Type 1 LVC

Thus, this grammar allows us to recognize the following patterns:

- On je *doveo u pitanje* …
- *U pitanje* je *doveo*…
- *Doveo* je u **veliko** *pitanje*…

A similar graph[3] exists for the recognition of Type 5 LVCs (Fig. 4). Since there is no preposition in this graph, there was no need to include the check for the case agreement. There is an additional subgraph named:afterN, that allows for some extra Genitive NPs to be recognized but only if the light verb is in inversion.

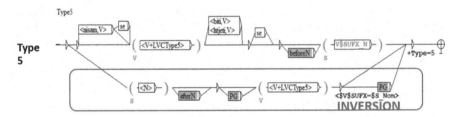

Fig. 4. Subgraph for the recognition and annotation of Type 5 LVC

Thus, for the entry *voditi borbu* (to lead a fight) <voditi,V+FXC+LVCType5 +SUFX=borba>, this grammar allows us to recognize the following patterns:

- On je *vodio borbu*…
- [NP insertion] On je *vodio* sve svoje *borbe*…
- [inversion] *Vodio* je *borbu*…
- [inversion + NP in Genitive insertion] *Borbu* stoljeća *vodili su*…

---

[3] The same logic present in graphs for the recognition of Types 1 and 5 is followed for the recognition of Types 2, 3 and 4.

7   LVC Grammar Results

The grammar was applied to three different subsections of Croatian corpus, each having the total number of words 23 111, 22 970 and 23 185, respectively. Each subsection has already been manually checked for LVCs (cf. Sect. 4). This allowed us to measure the recall as well as the precision of our grammar.

Since we have marked each recognized string with the type (we used the type of an LVC that the string belongs to), we were able to measure precision, recall and f-measure per type next to an overall performance of our proposed grammar (Table 2).

Table 2. Precision, recall and f-measure for each LVC type and an overall grammar performance

Type	Precision	Recall	F-measure
1 (100)	0.99	1	0.995
2 (54)	1	1	1
3 (4)	1	1	1
4 (3)	1	1	1
5 (369)	0.94	1	0.97
Overall	0.96	1	0.98

The preliminary results show satisfactory values. We were able to recognize different syntactic and morphological variations as can be seen from our concordance:

- *<LVC+Type=1>*

  - *<biti u padu>* Krijumčarenje narkotika *u padu je.*
  - *<uvrstiti na popis>* Tri filma iz jugoistočne Europe *uvrštena su na popis* 12 filmova koji će dobiti financijsku pomoć.
  - *<doći do napretka>* u Bugarskoj *došlo do određenog napretka* od objavljivanja

- *<LVC+Type=2>*

  - *<biti riječi o>* Bila je riječ o jednoj minuti za mir i…
  - *<biti riječi o>* Bilo je riječi o raznim infrastrukturnim investicijskim …
  - *<naći put do>* Verzija izvješća koja je procurila *našla je svoj put do* medija…
  - *<imati utjecaj na>* financijska kriza *imala relativno neznatan utjecaj na* Kosovo.
  - *<imati volju za>* sadašnja vlada *ima potrebnu političku volju za* što skorije ispunjavanje

- *<LVC+Type=3>*

  - *<biti u skladu s>* uz pojašnjenje da *nije bila u skladu s* financijskim planovima…
  - *<poslužiti kao temelj za>* okončao rat i *poslužio kao temelj za* mir, morao je…
  - *<biti u skladu s>* za koje je ocijenjeno da *su u skladu s* međunarodnim standardima…

- *<LVC+Type=4>*

  - *<postavljati uvjet za povlačenje>* plan misije, *postavljajući uvjete za postupno povlačenje…*
  - *<postojati izgledi za rješenje>* još uvijek *postoje mali izgledi za rješenje*
  - *<izdati nalog za uhićenje >* Sudovi su *izdali nekoliko naloga za uhićenje*

- *<LVC+Type=5>*

  - *<donijeti odluku>* vlada je *donijela ovu odluku* prije predstojećih rasprava…
  - *<izraziti potporu>* čija je zemlja *izrazila snažnu potporu* Ahtisaarijevu planu…
  - *<potpisati sporazum>* *Sporazum je potpisan* u Nicosiji.

There is still some fine-tuning that may be done to the grammar in order to improve the precision and lower the number of false positives. For example, when systems falsely recognize an LVC because the adjective and the verb have the same form as in this sentence:

- Američke Države i Rusija <u>predstavili</u> <u>su</u> *suprotstavljena* <u>stajališta</u>.

The grammar recognizes the string *su suprotstavljena stajališta* instead of the string *predstavili su stajališta*. This happened since there is an LVC entry '*suprotstaviti stajališta*', but also an entry '*predstaviti stajališta*'. However, in this example, the form '*suprotstavljena*' is an adjective to the noun '*stajališta*' and not a verb, i.e. passive verbal adjective of a verb '*suprotstaviti*'.

8 Conclusion

In this paper, we have presented the construction of NooJ dictionary and grammar used for the recognition of light verb constructions in Croatian. The grammar designed for this purpose was based on the previously developed database consisting of LVCs extracted from the UD Treebank for Croatian. However, having in mind the power of NooJ engine and its capabilities, we conducted a thorough analysis of LVCs and divided them into five major subtypes depending on the morphosyntactic properties of their nominal elements. Each property was coded into the LVC dictionary. This enabled us to reuse that information into the graphs that were designed in such a manner as to recognize light verbs that are syntagmatically preceded but also preceded by their nominal elements, i.e. when they appear in inversion.

In the future works, we plan to develop additional grammars for the recognition of LVSs, but with a slightly different objective. Namely, our aim will be to construct grammars that would be able to recognize light verbs that co-occur with particular nominal elements. In other words, we wish to recognize the full span of light verbs that co-occur with particular NPs or PPs in corpora.

References

1. Agić, Ž., Ljubešić, N.: Universal dependencies for Croatian (that work for Serbian, too). In: Proceedings of the 5th Workshop on Balto-Slavic Natural Language Processing, Hissar, Bulgaria, pp. 1–8 (2015)
2. Baldwin, T., Kim, S.N.: Multiword expressions. In: Indurkhya, N., Damerau, F.J. (eds.) Handbook of Natural Language Processing, 2nd edn, pp. 267–292. CRC Press, Boca Raton (2010)
3. Kocijan, K., Librenjak, S.: Recognizing verb-based Croatian idiomatic MWUs. In: Okrut, T., Hetsevich, Y., Silberztein, M., Stanislavenka, H. (eds.) NooJ 2015. CCIS, vol. 607, pp. 96–106. Springer, Cham (2016). https://doi.org/10.1007/978-3-319-42471-2_9
4. Gulić, A.: Klasifikacija perifraznih glagola u hrvatskom jeziku. M.A. thesis. Faculty of Humanities and Social Sciences, University of Zagreb (2015)
5. Katunar, D., Srebačić, M., Raffaelli, I., Šojat, K.: Arguments for phrasal verbs in Croatian and their influence on semantic relations in Croatian WordNet. In: Proceedings of the Eight International Conferences on Language Resources and Evaluation (2012)
6. Menac, A., Fink Arsovski, Ž., Venturin, R.: Hrvatski frazeološki rječnik. Naklada Ljevak, Zagreb (2014)
7. Rosén, V., et al.: MWEs in treebanks: from survey to guidelines. In: Proceedings of the 10th International Conference on Language Resources and Evaluation, Portorož, Slovenia, pp. 179–193 (2016)
8. Sag, I.A., Baldwin, T., Bond, F., Copestake, A., Flickinger, D.: Multiword expressions: a pain in the neck for NLP. In: Gelbukh, A. (ed.) CICLing 2002. LNCS, vol. 2276, pp. 1–15. Springer, Heidelberg (2002). https://doi.org/10.1007/3-540-45715-1_1
9. Silberztein, M.: Formalizing Natural Languages: The NooJ Approach. Wiley, USA (2016)
10. Silić, J., Pranjković, I.: Gramatika hrvatskoga jezika – za gimnazije i visoka učilišta. Školska knjiga, Zagreb (2005)
11. Šojat, K., Filko, M., Farkaš, D.: Verbal multiword expressions in Croatian. In: Proceedings of the Second International Conference Computational Linguistics in Bulgaria. Institute for Bulgarian Language, Bulgarian Academy of Science, pp. 78–85 (2016)
12. Vučković, K.: Model parsera za hrvatski jezik. Ph.D. dissertation. Faculty of Humanities and Social Sciences, Zagreb (2009)
13. Vučković, K., Tadić, M., Bekavac, B.: Croatian language resources for NooJ. CIT. J. Comput. Inf. Technol. **18**, 295–301 (2010)

Semantic Predicates in the Business Language

Maddalena della Volpe[1]([✉]), Annibale Elia[2], and Francesca Esposito[2]

[1] Department of Business, Management and Innovation System,
University of Salerno, Fisciano, Italy
mdellavolpe@unisa.it
[2] Department of Political, Social and Communication Sciences,
University of Salerno, Fisciano, Italy
{elia, fraesposito}@unisa.it

Abstract. In recent years, the interest in the use of language for business has grown. It is recognized that the hidden persuasive linguistic potential improves the company's positioning in the public consciousness. The language of the business world is multifarious: we try to identify its features and behaviour, considering the evolution that it has faced primarily with the globalization of markets. Business activities are so complex that they require the application of several disciplines at the same time and therefore the use of specific languages and technical terminology. In order to reach an efficient analysis of business language, this study explores the role of semantic predicates constructed from lexical and the syntactic structures in which they are placed within business communication contexts. From the point of view of LG framework, a set of lexical-syntactic structures defines the value of semantic predicates, while the arguments selected by each semantic predicate are given the value of actants, subjects included. The features of each verb are expressed by the application of the rules of co-occurrence and selection restriction, through which verbs select semantically their arguments to construct acceptable simple sentences. In this way, the entries belonging to electronic dictionaries should be classified presuming their similarity and proximity. Even if the list of semantic tags is not simply identifiable, grammars could be built for single sets of semantic predicates. LG descriptions assign correlated predicates and arguments by applying electronic dictionaries of Italian. Using NooJ environment and Italian linguistic resources to automatically processing natural language, we will process a corpus of business documents. We will show and describe the syntactic structures, semantic and syntactic properties of predicates, in order to build formal grammar for business language.

Keywords: Semantic predicates · Business language
Natural language processing · Text Mining · NooJ application

1 Introduction

In the business language, many special expressions are used to define and describe the actions of a company within itself or with the outside world. There is often only one verb to express a context, a process or an action. In this article, we will analyse the most common predicates in business documents to understand their functions and features.

S. Mbarki et al. (Eds.): NooJ 2017, CCIS 811, pp. 108–116, 2018.
https://doi.org/10.1007/978-3-319-73420-0_9

Moreover, through an automatic linguistic analysis, it is possible to verify the influence of the co-occurrences of these predicates in order to understand the text. By adopting the bipartition between operators and arguments, first proposed by Harris [1, 2], and subsequently adopted by Gross [3–9], we can assert that predicates assume the ability to request specific arguments and constitute a potential simple sentence. Although not verb-centric, Lexicon-Grammar, containing all the possible combinations of simple sentences at the distribution level, considers semantic information that allows us to recognize the predicates and their topics (operators) at a semantic level [10]. From the point of view of LG framework, a set of lexical-syntactic structures defines the value of semantic predicates, while the arguments selected by each semantic predicate are given the value of actants, subjects included [11]. The features of each verb are expressed by the application of the rules of co-occurrence and selection restriction, through which verbs select semantically their arguments to construct acceptable simple sentences. According to Monteleone and Vietri [12], we have semantic predicates expressing the intuitive notion of "exchange" (Transfer Predicates), "motion" (Movement Predicates) or production (Creation Predicates). Each set of semantic predicates assumes those arguments with which they have compatible semantic roles.

Transfer Predicates have a "giver", an "object to transfer" and a "receiver", as in the sentences:

1. Mario (giver) gave a cake (object to transfer) to Juliet (receiver)
2. Juliet (receiver) received a cake (object to transfer) from Mario (giver).

Movement Predicates select an "agent of motion", "object to move" and a "locative name", as in the following:

1. Mario (agent of motion) went to Paris (locative name)
2. Mario (agent of motion) moved the books (object to move) from his house to the office (locative names).

Creation Predicates, finally assume a "creator" and a "creation":

1. Mario (creator) wrote a novel (creation)
2. Juliet (creator) composed a song (creation).

In this way, the entries belonging to electronic dictionaries should be classified presuming their similarity and proximity to semantic predicates. Even if the list of semantic tags is not simply identifiable, due to the polysemy of simple nouns, grammars could be also built for single sets of semantic predicates. LG descriptions assign correlated predicates and arguments by applying electronic dictionaries of Italian. It is also possible to build grammars that annotate all specific semantic predicates. In the following paragraphs, we analyse the language of business, identifying features and singularity. Business documents define the complex world of enterprise and describe business activities, functions and actors. We analyse business plans to explore the use of language, and in particular in this study we focus on semantic predicates. According to Elia [10], we take into consideration LEG-Semantic Role Labelling system (LEG-SRL) for Italian, built on 2000 verbal uses, included in semantic predicates classes. We recognize some verbal uses recurring in the documents, with the purpose of explaining and improving the companies express themselves relying on communication exercise.

2   The Language of Business

As far as the business language is concerned, we must consider two fundamental aspects that make the analysis rather complex. To express business activities in their complexity, as well as in their diversity, we have to consider, on one hand, the sublanguages that characterize this world, and on the other hand terminology. For instance, sublanguages are used to describe professional activities belonging to different business sectors: banking, trading, accounting, communication, logistics, administration etc. Another issue is referred as terminology: no one could say that business has a specific and limited vocabulary. The study of language in business contexts is highly interdisciplinary [13]. Business activities are so complex that they require the application of several disciplines at the same time, and therefore the use of specific languages. Although, it is always necessary that the circumstances, in which terms are uttered, should be in some way, or ways, appropriate. The combination of business functions and processes is impacted by improved communication. From company to company, we have seen language skills consistently deliver tangible business value and virtuous results for organizations that invest in language training.

Ford and Wang [14] observed how the use of language in the field of strategic management has been the subject of many studies [15–18] just because there is no unique classification of words as it exists for other disciplines such as Economy. Every strategic document is a stream of decisions [19] and actions whereby it does not just describe reality but performs it in the same moment in which they are representing it.

The language of the business world is definitely multifarious: we have tried to identify its features and behaviour, considering the evolution that it has faced primarily with the globalization of markets. In the last thirty years, the interest of researchers in the variety of specific language uses has increased significantly [20]. However, in relation to different specialized varieties of the language, there is no a unified terminology, and tags used in this field of research by various researchers are different. Nevertheless, we must consider the fact that the use of certain terms entered in the common language through mass media, as we know, often becomes the point of contact between the specialists and the people. Thus, we will have a kind of coded language that is typically used in the field of the economy, and another type of language that instead has developed among the experts, a type of jargon, which then became part of everyday life through the media.

For instance, in previous studies [21, 22] we dealt with the specific lexicon used by media to describe the phenomenon of startup companies. We studied how the Italian terminology and this specialty language can be used in routine automatic text analysis. Using NooJ environment [23–28] for the automatic processing of natural language, through the application of electronic dictionaries of terminology and specialty, we analysed a corpus of 2000 journal texts centred on the startups topics. After the analysis, we collected about 400 entries, a great part of which belongs to the semantic fields of economics and informatics and a small part to professionals, revenue and law. Moreover, it appears that the terminology of the world of startups is rich of foreign words, coming mainly from the United States. Through the study of the presence, frequency and origin of lexical entries, it is possible to grasp certain phenomena implicitly expressed in the texts analyzed, with the objective of a better understanding of the evolution of the ecosystem of startups. The specialty language that has been determined requires the continuous, online monitoring of a dynamic and innovative vision in the specialized terminology field. On the other hand, it derives from the fact that there is a very strong presence of borrowings in the English language in the lexicon of startups. This data could be taken as an invitation to extend the research by adapting these terms to the Italian language system, thus satisfying the need to find effective correspondents to describe certain concepts. This case shows even how mostly technical words enter on our common language through mass media, and become our opportunity to comment some socio-economic events.

Nowadays, in the language of business, we can identify two level of language: specialized and popular. The specialty language includes all the features of the sector language, while the popular language is spread through mass media. The popular level resorted to some mitigations, making less complex the language, or recourse to metaphors. Predominantly, economic dictionaries characterize the language of Business, but the enterprise system is so complex that it naturally requires the intervention of more specialized languages in the interaction processes, based on the nature of the enterprise and on the market in which it operates. The recognition of economic terminology is revealed only as the basis for a larger study that may involve other types of specialized language processing, within the analysis of textual documents that provide information to support strategic decisions. Thus, the language of business is partially the language of Economics, as it uses many words that have a dramatic nuance ("crisieconomica" as "economic crisis") or military origin ("manovra finanziaria" as "budgetary manoeuvre") as shown by Parantainen [29]. The most striking feature of the business language in Italian is the presence of foreign words and expressions, especially of English origin, so abounding of technicalities and terms that are often incomprehensible to the experts. To obtain an efficient Text Mining system and to apply it to the business document analysis, we have to consider typical economic language, opening our analysis' field to other knowledge domains.

Business documents are files that provide details related to a company, in fact, they are used to communicate, transact business and analyse productivity. In the meantime, business documents provide the profile of an organization and may be referred to for years to come: it is very important that they are well prepared, to avoid conveying a negative impression about the person who wrote it or the company for which it is written. Thus, writing excellent business documents is imperative for any working professional: they can be digital, occurring as electronic files, or in a physical form, written or printed on paper. Business documents range from brief email messages to complex legal agreements. Some documents are prepared by employees and business owners, while others are drafted by professionals from outside the company, such as accountants and lawyers. The most important external and internal business documents are:

(a) Business plans
(b) Letters, mail and memorandum
(c) Business reports
(d) Financial and accounting documents
(e) Operational documents
(f) Customer documents.

After choosing documents' types that we would process, we proceed with pre-processing of unstructured linguistic data. This phase goes through the application of LG theory and methodologies formalization of language (LG tables, electronic dictionaries and local grammars). Then we can process the texts in NLP software environment. After this linguistic pre-processing phase, we obtain several results, which can be integrated into different business applications [30].

3 Semantic Predicates and Syntactic Structures Groupings

At this point, we provide an example of business document automatic analysis based on LG framework: we analyse a corpus of Business Plans, recognizing a set of semantic predicates used in business language. Subsequently, we create New Local Grammars and other tools, developing a complex system that allows understanding the features of language used by expert in this field. Corpus exploration in Fig. 1, leads us to recognize a substantial number of operators that present these arguments and that provide indications of circumstantial nature.

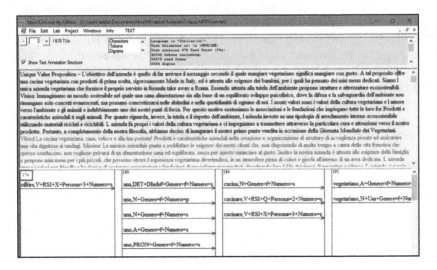

Fig. 1. Corpus exploration

The most frequent predicates in the corpus (about 100 Kb) are 21. We excluded from our observations the verbs that play only a supporting function for other verbs (to be or to have). We noticed that often some verbs are interchangeable with each other, in the sense that they have the ability to select the same lexical material that may co-occur with them, so we grouped them according to their behaviour (Table 1).

We can observe that as for the mentioned verbs, the action typically passes directly from the enterprise to the object (person, animal or thing) that receives or suffers it. Depending on their behaviour, we could associate these predicates to the grammatical classes already recognized in some previous studies [10]. Following the same examples taken from our corpus (Tables 2, 3, 4 and 5).

Table 1. Semantic predicates groupings

Transfer predicates N0 V N1 a N2	Offrire, vendere, distribuire, commerciare
Causative locative predicates N0 V N1 Loc N2	Posizionare, immettere, inserire, Introdurre
Communication predicates N0 V a N1	Proporre, presentare, garantire, assicurare
Creative predicates N0 V N1 = -um	Sviluppare, accrescere, espandere, potenziare, incrementare

Table 2. Transfer predicates (N0 V N1 a N2)

Agent giver	*L'azienda* offre strutture ecosostenibili agli ospiti
	The company offers environmentally friendly facilities to guests
Object of transfer	L'azienda offre *strutture ecosostenibili* agli ospiti
	The company offers *environmentally friendly facilities* to guests
Benef./receiver	L'azienda offre strutture ecosostenibili *agli ospiti*
	The company offers environmentally friendly facilities **to guests**

Table 3. Communication predicates (N0 V a N1)

Agent issuer	*L'azienda* garantisce la massima genuinità dei prodotti al cliente
	The company guarantees the highest genuineness of the products to the customer
Topic/message	L'azienda garantisce *la massima genuinità* dei prodotti al cliente
	The company guarantees *the highest genuineness* of the products to the customer
Benef./ *receiver*	L'azienda garantisce la massima genuinità dei prodotti *al cliente*
	The company guarantees the highest genuineness of the products *to the customer*

Table 4. Causative locative predicates (N0 V N1 Loc N2)

Agent	*L'azienda* immette un prodotto innovativo nel mercato
	The company introduces an innovative product into the
Place	L'azienda immette un prodotto innovativo *nel mercato*
	The company introduces an innovative product *into the market*

Table 5. Creative predicates (N0 V N1 = -um)

Agent creator	*L'azienda* accresce il fatturato
	The company increases the turnover
Topic/obj. of creation	The company increases the turnover
	L'azienda accresce *il fatturato*

The examples presented here are only classification principles, but such verbs with their uses, appear extremely frequent in the business language. We have created a local grammar on the basis of the most frequent simple sentence form in Italian [31] that we have found in the corpus. Some examples of local grammars are represented for transfer predicates in Fig. 2, communicative predicates in Fig. 3 and causative locative predicates Fig. 4. By conducting some experiments with NooJ, it is possible to label predicates and arguments to question the machine, with respect to the nature of the attendant, and of the main themes.

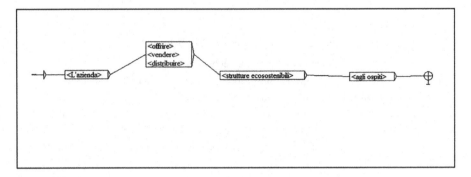

Fig. 2. Example of local grammar with transfer predicates

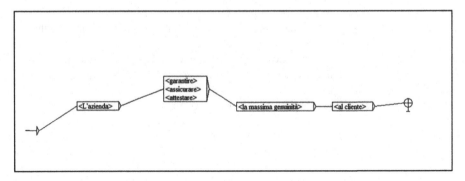

Fig. 3. Example of local grammar with communication predicates

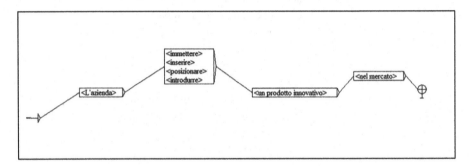

Fig. 4. Example of local grammar with causative locative predicates

4   Conclusions

To be competitive in the market and face innovation challenges, companies need to acquire specific knowledge, growing and communicating outside their values. Despite the detailed level of the methodological and theoretical framework provided, which have given us great hopes: the analysis of results has made us realise that the formalization of all linguistic phenomena is extremely complex. As we have tried to show in this paper, the exhaustive description of the lexicon and grammatical uses of a language, associated with a morphosyntactic electronic dictionary and a variety of local grammars could give satisfying results at this primary level. Semantic predicates could be used to analyse business processes, arguing that content of a text is unlabelled in advance, such as business plans, emails, and business formal communications. We admit that this study is a primary attempt to the development of a linguistic support to embed inside decision-making, with a particular reference to the document-driven analysis.

References

1. Harris, Z.S.: Distributional structure. WORD **10**, 146–162. Reprinted in Fodor, J., Katz, J.: The Structure of Language: Readings in the Philosophy of Language. Prentice-hall, Upper Saddle River (1964)
2. Harris, Z.S.: A Theory of Language and Information: A Mathematical Approach. Clarendon Press, Oxford, New York (1991)
3. Gross, M.: L'emploi des modèles en linguistique. Langages 9, pp. 3–8. Larousse, Paris (1968)
4. Gross, M.: Méthodes en syntaxe. Hermann, Paris (1975)
5. Gross, M.: Mathematical Models of Language. Prentice-Hall, Englewood Cliffs (1972)
6. Gross, M.: Méthodes en syntaxe, régime des constructions complétives. Hermann, Paris (1975)
7. Gross, M.: Les bases empiriques de la notion de prédicat sémantique. Langages, 63. Larousse, Paris (1981)
8. Gross, M.: Lexicon-Grammar. The Representation of Compound Words. In: AA. VV., Proceedings of COLING-1986, pp. 1–6. University of Bonn, Bonn (1986)
9. Gross, M.: La construction de dictionnaires électroniques, dans AA. VV. Ann. des Télécommun. **44**(1), 4–19 (1989). CNET: Issy-les-Moulineaux/Lannion
10. Elia, A.: Operatori, argomenti e il sistema "LEG-Semantic Role Labelling" dell'italiano. In: Relazioni irresistibili Pisa, pp. 105–118. ETS (2014)
11. Elia, A., Vietri, S., Monteleone, M., Marano, F.: Data mining modular software system. In: SWWS2010 – Proceedings of the 2010 International Conference on Semantic Web & Web Services, Las Vegas, Nevada, USA, 12–15 July 2010, pp. 127–133. CSREA Press (2010)
12. Vietri, S., Monteleone, M.: The NooJ english dictionary. In: Formalising Natural Languages with NooJ 2013: Selected Papers from the NooJ 2013 International Conference 12 Back Chapman Street, Newcastle upon Tyne, NE6 2XX, pp. 69–86. Cambridge Scholars Publishing (2014)
13. Studer, P.: Linguistics applied to business contexts: an interview with Patrick Studer. ReVEL **11**(21), 187–202 (2013)

14. Ford, E.W., Wang, Z.: Tackling the confusing words of strategy: effective use of key words for publication impact. Bus. Manag. Strategy, **5**(1) (2014)
15. Hoskisson, R.E., Hitt, M.A., Wan, W.P., Yiu, D.: Theory and research in strategic management: swings of a pendulum. J. Manag. **25**(3), 417–456 (1999)
16. Leontiades, M.: The confusing words of business policy. Acad. Manag. Rev. **7**(1), 45–48 (1982)
17. Nicolai, A.T., Dautwiz, J.M.: Fuzziness in action: what consequences has the linguistic ambiguity of the core competence concept for organizational usage? Br. J. Manag. **21**, 874–888 (2009). https://doi.org/10.1111/j.1467-8551.2009.00662.x
18. Ronda-Pupo, G.A., Guerras-Martin, L.Á.: Dynamics of the evolution of the strategy concept 1962–2008: a coword analysis. Strategic Manag. J. **33**, 162–188 (2012). https://doi.org/10.1002/smj.948
19. Mintzberg, H.: Patterns in strategy formation. Manag. Sci. **24**(9), 934–948 (1978)
20. Elia, A., Monteleone, M., Esposito, F.: Dictionnaires électroniques et lexique des startups. Un exemple d'analyse textuelle automatique. Dictionnaires électroniques et dictionnaires en ligne, Les Cahiers du dictionnaire **6**, 43–62 (2014)
21. Esposito, F., della Volpe, M.: Using text mining and natural language processing to support business decision: towards a NooJ application. In: Barone, L., Monteleone, M., Silberztein, M. (eds.) NooJ 2016. CCIS, vol. 667, pp. 234–245. Springer, Cham (2016). https://doi.org/10.1007/978-3-319-55002-2_20
22. Esposito, F., Elia, A.: NooJ local grammars for innovative startup language. In: Barone, L., Monteleone, M., Silberztein, M. (eds.) NooJ 2016. CCIS, vol. 667, pp. 64–73. Springer, Cham (2016). https://doi.org/10.1007/978-3-319-55002-2_6
23. Silberztein, M.: Nooj Manual (2003). http://www.nooj4nlp.net/NooJManual.pdf
24. Silberztein, M.: Corpus linguistics and semantic desambiguation. In: Maiello, G., Pellegrino, R. (eds.) Database, Corpora, Insegnamenti Linguistici. Linguistica n° 63, Schena Editore/Alain Baudry et C.ie, pp. 397–410 (2012)
25. Silberztein, M.: NooJ computational devices. In: Koeva, S., Mesfar, S., Silberztein, M. (eds.) Formalising Natural Languages with NooJ ·2013: Selected Papers from the NooJ 2013 International Conference (Saarbrucken, Germany), pp. 01–14. Cambridge Scholars Publishing, Newcastle (2013)
26. Silberztein, M.: NooJ V4. In: Koeva, S., Mesfar, S., Silberztein, M. (eds.) Formalising Natural Languages with NooJ 2013: Selected Papers from the NooJ 2013 International Conference (Saarbrucken, Germany), pp. 01–12. Cambridge Scholars Publishing, Newcastle (2014)
27. Silberztein, M.: Analyse et generation transformationnelle avec NooJ. In: Elia, A., Iacobini, C., Voghera, M. (eds.) 2015 Proceedings of the 47th Annual Meeting of the Italian Linguistic Society "Livelli di Analisi e Fenomeni di Interfaccia". Bulzoni, Rome (2015)
28. Silberztein, M.: La formalisation des langues: l'approche de NooJ. ISTE Ed, London (2015)
29. Parantainen, P.: I prestiti non adattati nel linguaggio dell'economia, Master's degree (2001). https://jyx.jyu.fi/dspace/handle/123456789/13653
30. Esposito, F.: Semantic Technologies for Business Decision Support. Discovering Meaning with NLP Applications, Ph.D. thesis (2017)
31. Vietri, S.: Dizionari elettronici e grammatiche a stati finiti. Metodi di analisiformaledella lingua italiana. Salerno, Plectica (2008)

Invalid Syntax: NooJ Assisted Automatic Detection of Errors in Auxiliaries and Past Participles in Italian

Ignazio Mauro Mirto[1(✉)] and Emanuele Cipolla[2]

[1] Università di Palermo, 90128 Palermo, Italy
ignaziomauro.mirto@unipa.it
[2] Consiglio Nazionale delle Ricerche, 90146 Palermo, Italy

Abstract. The work targets two areas of Italian morphosyntax: auxiliary selection (AS) and past participle agreement (PPA). In selecting such inflectional morphemes, learners of Italian commit frequent errors, even after a long period of constant study. We aim to enclose AS and PPA within the boundaries of NLP in order that a tool can be developed with a twofold purpose: first, it helps experts to build specific computer drills regarding AS and PPA; second, it assists self-taught learners in verifying whether their periphrastic sentences in Italian are well-turned. This area of Computer-Assisted Language Learning is currently poorly investigated. Further research might substantiate the importance of a field which is aimed at facilitating the study of a foreign language and simultaneously stimulate advances in NLP.

Keywords: Automatic identification of grammatical relations
Inflectional morpheme generation · CALL

1 Introduction

The authentic utterance in (1), produced by a native speaker of French with a forty year experience of Italian, is ill-formed:

(1) * Il numero di telefono ha cambiato 'The telephone number has changed'

The well-formed version of (1) is *Il numero di telefono è cambiato*, since in its intransitive use the verb *cambiare* is unaccusative[1]. This category of error[2] thus is in

---

The authors thank the Salerno team, in particular Annibale Elia, for sharing important material from their NooJ resources. Italian regulations on authorship require the authors to specify the sections for which they are responsible: EC is responsible for Sects. 3 and 4.2, whilst the remaining sections were contributed by IMM.

---

[1] It is beyond the scope of this paper to illustrate the difference between unergative and unaccusative predicates. In Relational Grammar, the framework which "saw" the birth of this finding, the distinction between the two types is purely syntactic (see [3]). It is common knowledge that this rule was originally proposed in [27].

[2] In regard to the well-known difference between 'mistake' and 'error', (see [4]).

S. Mbarki et al. (Eds.): NooJ 2017, CCIS 811, pp. 117–129, 2018.
https://doi.org/10.1007/978-3-319-73420-0_10

the area known as auxiliary selection (henceforth AS, [1, 2]). In Italian, AS works in part differently from French, which sheds light on the error committed in (1). To the best of our knowledge, NLP has paid only little attention to such challenges concerning foreign language teaching (see [5] on Computer-Assisted Language Learning or CALL[3]). It is the opinion of the authors of this paper that further investigation in this area might instead (a) substantiate the importance of a field of research which is aimed at facilitating the study of a foreign language and (b) simultaneously stimulate advances in NLP.

Learners of Italian also have great difficulty in the past participle agreement (henceforth PPA), another central morphosyntactic area of the language. An additional unacceptable utterance, produced by a female student[4] and reported in (2), exemplifies such difficulties, inasmuch as the past participle should agree with the subject and thus be *arrivata*, singular and feminine[5]:

(2) * Io   arrivato   in Libia   'I arrived in Libya'
     1SG arrived.MS.SG in Libya

The aim of this paper is to enclose the aforementioned grammatical areas within the boundaries of NLP in order that a tool can be developed with a twofold purpose: on the one hand, it helps experts to build specific computer drills regarding AS and PPA, and, on the other, assist self-taught learners in verifying whether their periphrastic sentences in Italian are well-formed. We will present a pilot study which calls for a number of non-trivial sub-tasks, including a tool for parsing a learner's simple sentence (i.e. with no subordinate clauses) and return an evaluation as regards the above-mentioned morphemes. A further aim is to develop an application for use on a smartphone.

2 Algorithms

Italian auxiliaries and past participles yield multiple intricacies which make their correct usage rather difficult. Grammars of standard Italian normally pay such matters close attention (see [6, 7] present a number of *putative* exceptions).

The reasons for focusing on AS and PPA derive from their formal treatment in the theoretical framework of Relational Grammar (henceforth RG). First, [1] (see also [2]) and then [8] succeeded in reducing the complexity of AS and PPA, respectively, to simple rules which were formulated as algorithms, a fact that paved the way for their use in NLP. Each rule comprises a default value (implicit in (4) below), as is often the case

---

[3] The CALL bibliography mainly concerns English [9, 10] points out that CALL is not a prominent area of computational linguistics.

[4] The recording is available at https://www.youtube.com/watch?v=hEiBNXdDRF8, 19:01 min.

[5] Sentence (2) is ill-formed also on account of the missing auxiliary. Italian being a pro-drop language, the correct sentence should be either *Sono arrivata in Libia* or *Io sono arrivata in Libia*.

with natural languages. Clearly, their interpretation and use require adequate knowledge of at least the fundamentals of the framework, which is, however, by and large easily acquired. The rules are formulated as follows (the PPA rule also works for adjectives)[6]:

(3) *Italian Auxiliary Selection* (adapted from [7])

An auxiliary is *essere* 'be' if the nominal heading its P-initial 1-relation also heads a 2-relation in the same clause. Otherwise it is *avere*.

(4) *Participle/Adjective Agreement In Italian* [7]

Let *p* be a participle/adjective bearing the P-relation in clause *b*. Then *p* inflects for gender and number iff:

(i) the P-final stratum of *p* is intransitive, and
(ii) a legal agreement controller exists.

Nominal *a* is a legal agreement controller iff it bears the 2-relation in clause *b*.

According to these rules, auxiliaries and PPA suffixes are both inflectional morphemes. This is the standard, traditional stance in relation to the latter type of morphemes, less so for the former. Since the very beginnings of the RG framework, inflectional morphemes have been regarded as cues which the string manifests, thereby allowing one to "see" the structure of the clause, i.e. to deduce the series of relationships holding between certain items of the clause (see [11]). Specifically, the rule in (3) regards the relational *career* of the final subject. A subject can also bear the direct object grammatical relation, as, for example, is the case with passives (here a given nominal bears the subject and the direct object relations in different strata) or in reflexives (the relations are then borne in the same stratum). If this is the case, the auxiliary is predicted as *essere* 'to be'. In other cases, the other value is selected and the auxiliary will be *avere* 'to have'. The auxiliary is thus taken as a variable, i.e. a morpheme that can take any of a given set of surface values (overall, two in AS, four in PPA). Notice that also the rule in (4) involves the 2-relation, i.e. direct objects, and that only a direct object can control PPA agreement, which in Italian is also sensitive to the transitive or intransitive status of a given stratum. PPA can thus be assimilated to the litmus paper test: if the string displays PPA, then the clause *must* contain, at some stratum, a direct object. The importance of this information for parsing operations in NLP cannot be underestimated.

Two examples will probably clarify the aforementioned rules better than a lengthy prose description. The sentences in (5) and (6) display an important difference concerning AS in Italian intransitive sentences:

---

[6] The 1-relation and the 2-relation stand for subject and direct object respectively. The P-relation is any predicative relation, regardless of which part of speech conveys it. RG is a multistratal theory, so that a single nominal can bear more than one grammatical relation in different strata. When a nominal simultaneously bears two grammatical relations in a single stratum, this is called 'multiattachment' (as in Table 8 below). The rules in (3) and (4) are parametrically conceived. With a number of differences, they produce AS (except for one type of unaccusative) and PPA also in French [7]. The error shown in (1) derives from the valence of *changer* in its intransitive use. Unlike its Italian counterpart, this verb combines with *avoir*: *Le numéro de téléphone a changé.*

(5) Lea è partita   'Lea left'   (6) Lea ha parlato   'Lea spoke'
     L.   is left.SG.FM   L.   has spoken.SG.MS

Sentence (5) shows an unaccusative pattern with *partire* 'to leave' (this verb behaves as the intransitive *cambiare* 'change' in (1) above), whilst sentence (6) displays an unergative pattern. The difference is clear: the auxiliary must be *essere* 'be' in (5), and *avere* 'have' in (6). The sentences behave differently also with regard to PPA: in (5), *partita* 'left' must agree with its feminine and singular subject, which means that this clause does contain a 2-relation, whilst in (6) there can be no agreement: the past participle takes the default value and we therefore know that at least one of the conditions in rule (4) is not met.

The structural differences between (5) and (6) can be comprehended at a glance by means of the corresponding relational diagrams, to which we can refer as «in an authentically mathematical fashion» (Rosen, class notes 1991). The relational structure of (5) is displayed in Table 1. If the RG framework is unfamiliar to the reader, such a diagram should probably be observed as shown in Table 2. Instead of the relevant morphemes, the bottom line of the diagram contains two variables in bold, namely X and Y, each corresponding to a segment computable by just looking at the three strata above. Table 2 can therefore be construed as a case of morpheme generation:

Table 1. The structure of sentence (5)

2		P	
1		P	
1		P	F
Lea		è	partita

By virtue of the algorithms in (3) and (4), the three strata structure easily permits the following computations: (a) whether the auxiliary will be *essere* or *avere*, and (b) whether the past participle will either agree with one nominal in gender and number or take the default value. Specifically, one should look at the leftmost column of Table 2, that of the subject (the final 1): since the subject is also a direct object (the 2-relation), due to the unaccusative valence of *partire*, the auxiliary will be *essere*. This is in contrast with the structure of (6): in the leftmost column of Table 3, no 2-relation is found simply because *parlare* 'to speak' is unergative. Consequently, the auxiliary will be *avere*:

Table 2. Morpheme generation with sentence (5)

2		P	
1		P	
1		P	F
Lea	X	partitY	

The past participle morpheme is also easily computed. Notice first that the diagram in Table 3 contains no 2-relation, which means that the past participle takes the default morpheme (singular and masculine). On the other hand, the diagrams for sentence (5) in Tables 1 and 2 do contain a 2 relation, and, in addition, the final stratum of the past participle (the gray strata make it easily visible) is intransitive: the two conditions expounded in rule (4) are met, and the nominal bearing the 2-relation controls number and gender agreement (singular and feminine).

Table 3. The structure of sentence (6)

1		P
1	P	F
Lea	ha	parlato

It is a paradox that linguists themselves are generally quite sceptical with regard to the efficacy of the rules which are designed to compute specific pieces of morphology, like auxiliaries and clitics. This might be due to their preference for non-formal theories or perceived difficulties in Chomskyan analyses, which in our opinion face one or more of the following challenges: (a) rapid changes in the framework (see [12]), (b) either marginal or substantial differences in the ways each scholar formulates a certain rule, or (c) partial coverage of the empirical data of the language analyzed (this is the case for the PPA rule for Italian proposed by [13], as [7]) have demonstrated). Predictably, such scepticism spreads to computer scientists.

As for the algorithms provided in (3) and (4), which were developed in the 1970s for AS, and in the 1980s for PPA, it can be safely stated that they have since remained practically unmodified. A number of the clause types for which they yield the correct structures have been illustrated by [7] and exemplified in Sect. 4.4.

Several attempts have been made to account for the distinct auxiliaries in unergatives and unaccusatives on semantic grounds (see [14]), albeit to no avail, as [3] has convincingly argued. Thus learners can only rely on their memory (as is the case with the masculine or feminine gender of nouns): to perform successfully, they have to know the valence of an intransitive predicate, e.g. whether it is unergative or unac-cusative. Specifically designed computer drills in the area of NLP could benefit learners, inasmuch as the rules above can help their training. Thus, after typing the text, a learner who wants to verify the correctness of the sentences below:

(7)   a. * Lea ha partito   b. * Lea ha partita   c. * Lea è partito
       d. * Lea è parlato   e. * Lea è parlata   f. * Lea ha parlata

Will soon know that none of them is correct. This learner can either keep trying or request the tool for the correct output. In this regard, a prototype virtual assistant will also be developed (it has already been subscribed to the Telegram instant messaging network at the following address: http://www.telegram.org/). This virtual assistant will solicit input sentences from subscribers and answer with detected errors, if any, or with paraphrases. This assistant will be accessible from any software platform supported by a Telegram client, including mobile devices and computers.

3 Previous Studies

In order to understand the state of the art in relation to the automatic identification of grammatical relations, we can refer to [15]: «A major and critical difference between earlier studies of automatic extraction of grammatical information [...] and recent approaches is the explicit exploitation of grammatical knowledge. Earlier works rely solely on stochastic information».

The task cannot be of topical interest if one considers that, in the renowned journal *Computational Linguistics*, the metalinguistic words *subject* and *object* have occurred in none of the article titles of at least the last ten years. To the best of our knowledge, there exist no rule-based attempts with which to extract grammatical relations. Even probabilistic programs are not easily found. A statistical parser is offered by [16], whose precision rate is given as 90% at the most [17]. A paper regarding subject/object identification in Italian is also stochastic. There are two programs in Italian with available online demos, i.e. *LinguA* (http://linguistic-annotation-tool.italianlp.it/), and an elaboration of *TextPro* (http://textpro.fbk.eu/), called *TINT* (http://simpatico.fbk.eu/demo/), again both probabilistic in nature. They perform comparatively well in relation to the tools available for other languages (English in particular), albeit with a number of recalcitrant errors also in relation to AS and PPA which render them unsuitable for our purposes.

4 Procedure

Parts of the pre-processing stage are taken for granted, for instance with regard to tokenization and word sense disambiguation (e.g. *visto*, meaning either '*seen* or *passport visa*', *partito*, meaning either '*left* (past participle of *leave*) or *political party*'). Moreover, no chunking is necessary: the user's input is parsed at the outset to verify the presence of stopping conditions, one of them being the presence of two or more lexical verbs. If this is the case, as for example in *Mi piace correre* 'I like to run', a sentence where the post-verbal infinitive bears the subject relation, the user receives information about the undue complexity of the sentence.

The use of the algorithms for AS and PPA requires a certain amount of "grammatical knowledge", as [15] have stated. First, of key importance is identifying which nominal in the sentence is the subject. Second, identifying which nominal, if any, is the direct object is also crucial. The automatic identification of such functions is not an easy task. In our opinion, in order to achieve this result it is necessary to make use of the argument structure of the lexical verb. For example, it has already been pointed out that the verb *cambiare* 'to change' has at least two valency frames, one being intransitive, more precisely unaccusative, as in the utterance in (1), whilst the other is transitive (e.g. *Max ha cambiato il numero di telefono* 'Max changed his telephone number'). This transitive/unaccusative correlation also regards a number of verbs in English, for example *break, accumulate, disperse, dissolve*. The tables developed in the Grossian *Lexique Grammaire* framework by the Salerno team could probably be exploited on this matter (for French, see [26]). The procedure consists of four main steps.

4.1   Step 1 - Stopping Conditions

A function prompts the parser in order to establish whether the string contains the auxiliary + past participle sequence (in this order only). If this is not the case, the user receives a message stating that the sentence does not belong to this particular domain. Step 1 presents a potential problem: for example, sequences such as *Leo è raffreddato* 'Leo has a cold' only contain the sequence on the surface, given that in this case *raffreddato* functions as an adjective. This can be easily verified by contrasting the sentence with *Leo si è raffreddato* 'Leo caught a cold': in the former, *raffreddato* can take the superlative suffix (*Leo è raffreddatissimo*); in the latter, it cannot (**Leo si è raffreddatissimo*). This problem need not be tackled further because the algorithm in (4) also works for adjectives. Of note is also the fact that two items at the most can occur between the auxiliary and the PP; both must belong to the adverbial category (as, for example, in *Leo ha molto rapidamente verificato tutto* 'Leo checked everything very rapidly'). Multi-word adverbials such as *di punto in bianco* 'suddenly' must be considered as a single item.

4.2   Step 2 - Extracting Syntactic and Semantic Features

An instance of a possible input string is provided in line 1 of Table 4 below. The sentence is *Mio cugino ieri ha telefonato alla segretaria* 'Yesterday my cousin called the secretary':

Table 4. PoS Tagging and syntactic/semantic features

1	*mio*	*cugino*	*ieri*	*ha*	*telefonato*	*alla*	*segretaria*
2	SG-POSS	NOUN	ADV	AUX	PP	PREP-ART	NOUN
3	1ST	3RD/+HUM	-	3RD	-		3RD/+HUM
4	MS	MS	-	-	MS	FM	FM
5	SG	SG	-	SG	SG	SG	SG
6	-	-	-	-	$[1, (3)]$	-	-

The PoS-Tagging is given in line 2; line 3 concerns the extraction of the syntactic and semantic features [person] and [animacy] respectively; line 4 concerns [gender], line 5 [number], and, finally, line 6 provides the valence of *telefonare* 'to telephone'. This unergative verb obligatorily licenses a 1-relation and optionally (as the brackets indicate) an indirect object invariably introduced by the preposition *a* 'to/at':

With the exception of line 6, this cluster of information can be retrieved from the NooJ platform, as exemplified in Fig. 1. In order to make the information above usable in the following phases, a prototype software library proved to be necessary.

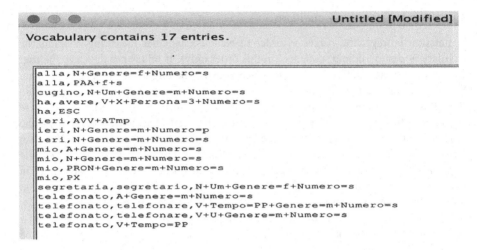

Fig. 1. An example of information retrieval with NooJ

4.3   Step 3 - Locating the Subject

A key point in the program is the search for the sentence subject. In Italian, a property of subjects is their being *noun* phrases. Prepositional phrases cannot fulfil this function (with a few exceptions: *In molti le hanno creduto* 'Many believed her'). With the elimination of prepositional phrases, the only 'survivors' in the string are mainly noun phrases[7], which either fulfil the subject or direct object functions. For example, in the sentence included in Step 2 above, the removal of the prepositional phrase *alla segretaria* 'to the secretary' would correctly leave the noun phrase *mio cugino* 'my cousin' as the only possible candidate for subjecthood. Thus, with intransitive verbs such as *telefonare*, the identification of the subject can prove to be a relatively simple task, whilst transitive verbs pose a greater difficulty. The module relating to locating the subject has been dealt with by [18]. They propose a rule-based script which makes use neither of Chomskyan-based trees nor of dependency parsing. Knowledge of the verbs' argument structure(s) is the key to concluding the task successfully. Of important use is also the delimiting of the verb phrase, including all clitics, which provides invaluable information concerning the argument structure of the deployed lexical verb.

A challenge which any NLP tool for Italian has to face in order to identify grammatical relations arises from the partitive article, whose paradigm contains the following forms: *del, dello, dell', della, dei, degli, delle*. This article is formed by mandatorily adding a form of the definite article to the preposition *di* (which therefore does not belong to the paradigm). The outcome is problematic for Italian NLP because this word-formation process gives rise to forms which are identical to those using *di* 'of' which are labelled as *preposizioni articolate*, namely 'prepositions endowed with an article'. That is, the identity of forms creates a challenge in the field of *Word Sense*

---

[7] Other segments may remain, for example adverbs (which are easily treated since they are not classified as (pro)nouns), compounds, and dates. See [18].

Disambiguation. For example, the sentence *Hanno parlato dei professori* is ambiguous in meaning: (a) 'They spoke *about* the professors', in which *dei professori* functions as a prepositional phrase, and (b) '*Some* professors spoke', in which *dei professori* functions as a noun phrase. Crucially, in the second meaning the phrase *dei professori* works as the sentence subject, that is a noun phrase, and is therefore the target of the search. This in turn means that the above mentioned function operating on prepositional phrases will eliminate all of them except the *preposizioni articolate* formed with *di*. The latter requires a further function (see [19]) which is able to disambiguate *about*-phrases (i.e. complements) versus other phrases, working as either the subject of the sentence, as in the example above, or the direct object, as in *Ho visto dei professori* 'I saw some professors'.

4.4 Step 4 - Relational Diagrams

The parsing procedure introduced above aims to obtain relational diagrams such as those illustrated in Tables 1 and 3. These syntactic configurations are constructed by extracting information from the string and therefore cluster the grammatical features in an easily retrievable way. This section will provide a few examples of the procedure followed to arrive at a few configurations and then apply the rules for AS and PPA.

The stopping conditions guarantee the diagram below as the starting point, since the pre-processing stage and Step 1 ensure (a) the presence of an auxiliary followed by a past participle, and (b) the absence of more than one lexical verb (Table 5):

Table 5. Starting diagram

	P
P	F
AUX	PP

The identification of the sentence subject leads to the development shown in Table 6. It is so because (a) all diagrams must contain a 1-relation, and (b) by definition the auxiliary inherits the 1-relation from the previous stratum:

Table 6. Unergatives

1		P
1	P	F
(pro)noun	AUX	PP

The diagram above is identical to the one illustrated in Table 3. It predicts that the AUX will be *avere* (because at no stratum does the final 1 bear the 2-relation) and that the PP will take the singular and masculine default value (because the diagram does not contain a 2-relation). Table 6 provides the *complete* configuration if an unergative verb licenses one argument only. Such a diagram is of particular significance because it is common to *all* periphrastic sentences. Any other diagram differs for additional strata, additional columns, or both (see below).

If the verb licenses only a direct object, then the configuration differs from that in Table 6 by only one additional stratum, the initial one, which is determined by the unaccusative valence:

Table 7. Unaccusatives

2		P
1		P
1	P	F
(pro)noun	AUX	PP

This is the configuration discussed in relation to sentence (5) above. The advancement from 2 to 1 (marked in gray) takes place because any configuration must contain a (final) 1-relation. The diagram yields the following values as regards the inflectional morphemes: the auxiliary will be *essere*, and the PP will agree with the nominal heading the 2-relation.

Direct reflexives such as *Lia si è derisa* 'Lia mocked herself' [7] differ from the unaccusative structure only by their initial stratum. They are characterized by the configuration provided below:

Table 8. Direct reflexives

1,2		P
1		P
1	P	F
(pro)noun	clitic + AUX	PP

In Table 8, the lexical verb is transitive. However, a single nominal bears the 1-relation and the 2-relation in the same stratum (it therefore carries two semantic roles). Once again, for the very same reasons provided above in commenting on Table 7, the configuration predicts the correct inflectional morphemes for AS and PPA.

Should a transitive verb present itself in the active voice, the diagram would then appear as in Table 9. The configuration generates the auxiliary *avere* and the default value for the PP because the condition (i) in rule (4) has not been met:

Table 9. Transitives

1		P	2
1	P	F	2
(pro)noun	AUX	PP	noun

The cases shown above are only a small part of the structures which the rules in (3) and (4) cover. The following sentences give a better idea of the number of clause types treated by [7] (the morphemes computed are in bold):

(8) Transitives with a direct object clitic (page 3)
     Eva *le ha* sperperate   'Eva squandered them'

(9) Adjectives/Serial verbs (page 4)
     Eva *era*/divenne bellissima   'Eva was/became very beautiful'

(10) Causative *rendere* 'make' (page 5)
     La situazione economica *ha* res*o* necessarie le misure restrittive
     'The economic situation has made the restrictive measures necessary'

(11) Passives (page 5)
     La rottura della tregua *è stat-a* condannat*a* dal Consiglio di Sicurezza
     'The violation of the truce has been censured by the Security Council'

(12) Impersonal sentences: Unspecified human subject (unergatives, page 6)
     Si è cenat*o* 'One dined'

(13) Impersonal sentences: Unspecified human subject (unaccusatives, page 6)
     Si è pervenut*i* all'accordo di sospendere le ostilità
     'One arrived at the agreement to suspend hostilities'

(14) Indirect reflexives (with two possible PP agreements, page 7)
     I senatori *si sono* concess*i/e* altre ricche prebende
     'The senators awarded themselves additional generous benefits'

(15) Antipassive (with two possible PP agreements, page 18)
     I Verdi *si sono* scordat*i/e* le chiavi   'The Verdis forgot their keys'

5 Concluding Remarks

In the renowned volume *Speech and Language Processing*, [20] dedicate a few lines to auxiliaries[8], which they define as «words […] that mark certain *semantic* features of a main verb» (2000: 294, our emphasis). The stance taken in this paper is different. In RG, auxiliaries are considered as purely syntactic items. They contribute verb-subject agreement, tense, etc., i.e. inflectional morphemes which are generated *by the clause*, not by the verb. That is, it is the clause type that selects the auxiliary, not the verb. In more technical terms, auxiliaries provide the 'syntactic finish' which is necessary to produce a finite clause [21]. Of crucial importance for this viewpoint is an empirical observation by [22], who has noticed that *all* Italian verbs can combine with *essere* 'to be'. Open to question is also the ontological foundation of [20]'s volume: «words are the fundamental building blocks of language» (2000: 19). Among NLP experts, this belief appears to be extremely widespread.

---

[8] No mention is made of PPA, most probably because the sample language is mainly English, whose past participles, as is well-known, do not inflect.

[20]'s approach, which can be summarized as 'semantics first', in which word meanings pre-exist the contexts of occurrence, is common also among linguists. Most probably, the automatic identification of grammatical relations, a key component of the program proposed above, does not even impinge on the consciousness of most computer scientists as a major theoretical problem in NLP. Nowadays, the prevailing fields are, for example, sentiment analysis and irony detection. A significant number of scholars concentrate on the so-called 'semantic web' and probably believe that complete knowledge of the meaning conveyed in a text can be achieved without paying attention to syntax.

The ontology on which we rely is along the lines of that of Maurice Gross and thus differs from that by [20]. Gross takes the simple clause as the smallest unit in language (see [23]), an approach which squares with Saussure's expectation: «Il y aura un jour un livre spécial et très intéressant à écrire sur le rôle du mot comme principal perturbateur de la science des mots» (as cited in [24]). We wonder whether the 'semantics first' approach correlates directly with the shortage of rule-based methods in NLP and the predominance of stochastic methods. Regarding the latter, we share [25]'s view: «even a "magical" statistical tool that could be used to build spectacular NLP applications but did not explain anything about the language is of little interest to us».

References

1. Perlmutter, D.: The Unaccusative hypothesis and multiattachment: Italian evidence. Paper Presented to the Harvard Linguistics Circle, 9 May 1978
2. Rosen, C.G.: The relational structure of reflexive clauses: evidence from Italian. Ph.D. dissertation, Harvard University (1981)
3. Rosen, C.G.: The interaction between semantic roles and initial grammatical relations. In: Perlmutter, D., Rosen, C. (eds.) Studies in Relational Grammar, pp. 38–77. University of Chicago Press, Chicago (1984)
4. Corder, S.P.: The significance of learners' errors. IRAL 5(4), 161–170 (1967)
5. Levy, M.: CALL: Context and Conceptualisation. Oxford University Press, Oxford (1997)
6. Lepschy, A.L., Lepschy, G.: The Italian Language Today. New Amsterdam Books, New York (1977)
7. La Fauci, N., Rosen C.G.: Past participle agreement in five Romance varieties. Texto! 15(3) (2010/1993)
8. La Fauci, N.: Oggetti e soggetti nella formazione della morfosintassi romanza. Giardini (Nuova collana di linguistica, 4), Pisa (1988)
9. Leacock, C., Chodorow, M., Gamon, M., Tetreault, J.: Automated grammatical error detected for language learners. In: Hirst, G. (ed.) Synthesis Lectures on Human Language Technologies, vol. 25. Morgan & Claypool (2014)
10. ten Hacken, P.: Computer-Assisted Language Learning and the Revolution in Computational Linguistics. http://www.linguistik-online.de/17_03/tenHacken.pdf. Accessed 01 Sept 2017
11. Gerdts, D.: Relational visibility. In: Dziwirek, K., et al. (eds.) Grammatical Relations: A Cross-Theoretical Perspective, pp. 199–214. Stanford University, CSLI, Stanford (1990)
12. La Fauci, N.: Relazioni e differenze. Questioni di linguistica razionale. Sellerio, Palermo (2011)
13. Burzio, L.: Italian Syntax: A Government-Binding Approach. Reidel, Boston, Dordrecht (1986)

14. Sorace, A.: Gradience at the lexicon-syntax interface: evidence from auxiliary selection and implications for unaccusativity. In: Alexiadou, A., et al. (eds.) The Unaccusativity Puzzle, pp. 243–268. Oxford University Press, Oxford (2004)

15. Huang, C.-R., Hong, J.-F, Ma, W.-Y, Šimon, P.: From corpus to grammar: automatic extraction of grammatical relations from annotated corpus. http://www.iis.sinica.edu.tw/papers/ma/19354-F.pdf. Accessed 09 Sept 2017

16. Carroll, J., Briscoe, T.: High precision extraction of grammatical relations. In: Bunt, H., Carroll, J., Satta, G. (eds.) New Developments in Parsing Technology, vol. 23, pp. 57–72. Springer, Dordrecht (2005). https://doi.org/10.1007/1-4020-2295-6_3

17. Dell'Orletta, F., Lenci, A., Montemagni, S., Pirrelli, V.: Climbing the path to grammar: a maximum entropy model of subject/object learning. In: Psychocomputational Models of Human Language Acquisition, pp. 72–81. Association for Computational Linguistics, New Brunswick (2005)

18. Mirto, I.M., Maisto, A.: Della identificazione automatica del soggetto nella proposizione semplice, paper submitted to the CLiC-it 2017 conference

19. Mirto, I.M., Cipolla, E.: Dalla word sense disambiguation alla sintassi, paper submitted to the CLiC-it 2017 conference

20. Jurafski, D., Martin, J.H.: Speech and Language Processing. Prentice Hall, Upper Saddle River (2000)

21. La Fauci, N., Mirto, I.M.: Fare. Elementi di sintassi. ETS, Pisa (2003)

22. La Fauci, N.: Compendio di sintassi italiana. Il Mulino, Bologna (2009)

23. Elia, A.: Operatori, argomenti e il sistema "LEG-Semantic Role Labelling" dell'italiano. In: Mirto, I.M. (ed.) Le relazioni irresistibili, pp. 105–118. Pisa, ETS (2014)

24. La Fauci, N.: Relazioni e differenze. Questioni di linguistica razionale. Sellerio, Palermo (2011)

25. Silberztein, M.: Formalizing Natural Languages. The NooJ Approach. ISTE Ltd., Wiley, London, Hoboken (2016)

26. Gardent, C., Guillaume, B., Perrier, G. Falk, I.: Maurice Gross' grammar lexicon and Natural Language Parsing. https://members.loria.fr/CGardent/publis/poznan05-synlex.pdf. Accessed 08 Sept 2017

27. Perlmutter, D.: Multiattachment and the unaccusative hypothesis: the perfect auxiliary in Italian. Probus **1**, 63–119 (1989)

28. Rosen, C.G., La Fauci, N.: Ragionare di grammatica. Un avviamento amichevole. ETS, Pisa (2017)

Some Aspects Concerning the Automatic Treatment of Adjectives and Adverbs in Spanish: A Pedagogical Application of the NooJ Platform

Andrea Rodrigo[1(✉)], Silvia Reyes[1], and Rodolfo Bonino[2]

[1] Facultad de Humanidades y Artes, Universidad Nacional de Rosario,
Rosario, Argentina
andreafrodrigo@yahoo.com.ar, sisureyes@gmail.com
[2] IES N° 28 "Olga Cossettini", Rosario, Argentina
rodolfobonino@yahoo.com.ar

Abstract. The purpose of this paper is to present some elements of analysis for the automatic treatment of adjectives and adverbs in Spanish with NooJ. The dictionaries of adjectives and adverbs of the Spanish Module Argentina are the result of combining the dictionaries designed by the research team Infosur led by Dr. Zulema Solana (Universidad Nacional de Rosario), with the Spanish Module Spain-UAB (Universidad Autónoma de Barcelona) supervised by Dr. Xavier Blanco. Following [1], we take as a reference the nucleus verb phrase, and assume there are nucleus adjective and adverb phrases as well. A syntactic unit is presented to integrate morphology and syntax in a preliminary way, and to analyse nucleus adjective phrases (SADJN), and nucleus adverb phrases (SADVN). Finally, an exercise thought of as a pedagogical application of NooJ for students to apply the grammar to a given text is proposed.

Keywords: NLP · NooJ · Pedagogy · Spanish adjectives · Spanish adverbs

1 Introduction

This paper is part of the research project "Computer tools in language processing: a pedagogical application for teacher training". The project has been accredited by the Ministry of Science, Technology and Innovative Production of the province of Santa Fe (Argentina), and is carried out by academics and researchers of two Argentine educational institutions: the Instituto de Enseñanza Superior IES N° 28 "Olga Cossettini" and the Universidad Nacional de Rosario.

To be able to work with Spanish adjectives in NooJ, the research group IES_UNR proceeded to adapt the dictionaries created by the research team Infosur, and to integrate the contributions of the Spanish Module Spain-UAB.

© Springer International Publishing AG 2018
S. Mbarki et al. (Eds.): NooJ 2017, CCIS 811, pp. 130–140, 2018.
https://doi.org/10.1007/978-3-319-73420-0_11

2 The Adjective in Spanish

In Spanish, adjectives show gender and number markings. They may be classified into different types according to different semantic or syntactic criteria. For instance, "apocopable" adjectives may be shortened by apocope in certain positions. Our interest lies in determining types of adjectives in relation to how they combine with other word categories, and to the position they occupy in a phrase. Our attention will be focused on the order changes allowed and on the available combinations.

As participles can be used as adjectives, they are grouped into the category of adjectives. Only the masculine singular (past) participle is used in the formation of compound tenses. But all participles can occur in adjective phrases, whether they are masculine, feminine, singular or plural. It should be also noted that some adjectives can have diminutive and superlative degrees.

2.1 The Spanish Adjective in NooJ

The Spanish Module Argentina dictionary of adjectives contains at the moment more than 18,000 entries. Several lexical entries and their inflectional paradigms will be shown below. Each lexical entry was randomly selected, for each represents the other members of its inflectional paradigm:

gordo, ADJ + FLX = FLACO
inconsecuente, ADJ + FLX = RECURRENTE

Participles are added at the end of the dictionary, and identified by the tag [ppio]:

compensado, ADJ + FLX = FLACO + ppio

As regards spelling variants, and taking as a reference the dictionaries of the Royal Academy of Spanish (RAE), each spelling is entered into a different entry, and if applicable, each variant is assigned a different inflectional paradigm.

antipiquete, ADJ + FLX = RECURRENTE
antipiquetes, ADJ + FLX = BURDEOS

If an adjective belongs to two inflectional paradigms, it is entered twice in the dictionary:

antigás, ADJ + FLX = BURDEOS
antigás, ADJ + FLX = COMÚN

2.2 Inflectional Paradigms

The inflectional paradigms assigned to most Spanish adjectives are shown below:

FLACO = <E>/masc+sg | s/masc+pl | <B> a/fem+sg | <B> as/fem+pl;

This paradigm comprises a great number of adjectives, and most adjectives inflect in this way. Within this group the adjectives tagged [+ ppio] are included.

RECURRENTE = <E>/_+sg | s/_+pl;

In this paradigm, adjectives like *inteligente, triunfante*, are included. The dash means that these adjectives agree with masculine as well as with feminine nouns. The majority of these adjectives possess the feature [+ noinicial].

ÁGIL = <E>/_+sg |es/_+pl;

It includes adjectives like *accidental, gremial*. The dash refers to masculine as well as to feminine nouns. A large number of these adjectives have the feature [+noinicial]

OPOSITOR = <E>/masc+sg | es/masc+pl | a/fem+sg | as/fem+pl;

This paradigm consists of adjectives like *aclarador, deshollinador, desgarrador*.

HARAGÁN = <E>/masc+sg | <A> es/masc+pl | <A> a/fem+sg | <A> as/fem+pl;

Adjectives like *albanés, besucón, bocón*, belong to this paradigm.

The inflectional paradigms for the Spanish adjective are thirteen, and for reasons of space, the paradigms that comprise few adjectives will not be discussed here.

To know the frequency of adjectives comprised in the dictionary in terms of the inflectional paradigms to which they belong, Standard Score is used. Saving the whole dictionary as a text, adjective frequency is analysed with Locate, and then, a statistical analysis of all occurrences is done (see Fig. 1).

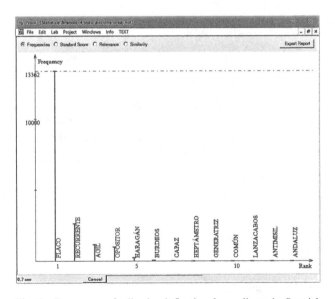

Fig. 1. Frequency of adjective inflectional paradigms in Spanish

2.3   Apocope and Order Constraints

Adjectives like *grande, bueno*, and *malo* are "apocopable", i.e. they are shortened by apocope when they precede a masculine singular noun. So, it is possible to say:

un buen hombre/un hombre bueno (a good man), but not **un bueno hombre*.
estos hombres buenos/estos buenos hombres (these good men)

As it can be clearly seen, constraints occur only in the singular. To mark these constraints, we use the features [apócope] and [apocopable]. Adjectives are entered in the dictionary with one of these two features:

buen, ADJ+masc+sg+apócope
bueno, ADJ+FLX=FLACO+apocopable

Likewise, there are a large number of adjectives that can never precede nouns. Some of them are relational adjectives, as the following:

sede judicial (judicial venue), but not **judicial sede*
rasgo bilabial (bilabial feature), but not **bilabial rasgo*

Many other adjectives, even though they are not relational, cannot either precede nouns. To all these adjectives, we add the feature [noinicial]:

republicano, ADJ+FLX=FLACO+noinicial

2.4   A Morphological Grammar

A morphological grammar to recognize the superlative degree of adjectives is created, specifically for the superlatives constructed with the suffix *-ísimo*. Superlatives ending in *-érrimo* are not included herein, for they are only a few, and are entered in the dictionary with the feature [superl]:

misérrimo, ADJ+FLX=FLACO+superl

This grammar can also recognize diminutives. It is a non-specific grammar able to recognize any sequence of letters, and suitable for adjectives and adverbs as well (see Fig. 2). The adverb will be discussed in the second part of our study.

When this grammar is applied to a specific sample of adjectives and adverbs, it can recognize both superlatives and diminutives. Being a non-specific grammar, it would not be appropriate to generate words, but it can indeed recognize superlative and diminutive degrees.

The analysis of *abrigadísimo*, the superlative degree of the adjective *abrigado*, is as follows:

abrigadísimo, ADJ+superl+masc+sg
abrigadísimo, ADV+superl

2.5   A Syntactic Grammar for the Nucleus Adjective Phrase

Following [1], it is assumed that the adjective forms a nucleus adjective phrase (SADJN) that can either be inside the nucleus noun phrase (SNN), or adjacent to it, as part of a greater unit, the noun phrase (SINNOM).

In this way, a nucleus noun phrase (SNN) is identified as follows: *[la casa]*, where the nucleus is at the end of the phrase, i.e. *casa*. In *[la [hermosa] casa]*, "the beautiful house," the adjective precedes the nucleus, and so the nucleus adjective phrase

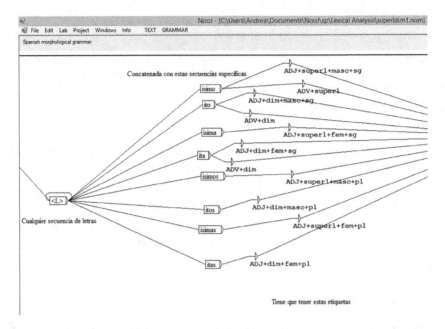

Fig. 2. A morphological grammar for adjectives and adverbs in Spanish

(SADJN), *hermosa*, is inside the nucleus noun phrase (SNN), *la casa*. On the other hand, in a noun phrase (SINNOM) like *[[la casa][hermosa]]*, "the beautiful house," *hermosa* is a nucleus adjective phrase (SADJN) inside the noun phrase (SINNOM), but adjacent to the nucleus noun phrase (SNN). The syntactic grammar recognizes different adjectives inside the noun phrase (SINNOM), or inside the nucleus noun phrase (SNN).

Some nucleus adjective phrases (SADJN) inside the SNN, e.g. *bellas*, en *[las [bellas] mañanas]*, are illustrated in the following graph, <ADJ+fem+pl-noinicial> (see Fig. 3):

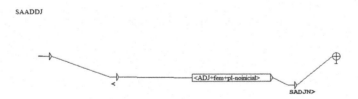

Fig. 3. Embedded graph for feminine plural adjectives inside the SNN

Adjectives shortened by apocope, like *buen*, in *[un [buen] hombre]* "a good man," are represented by the following graph, <ADJ+masc+sg-apócope-apocopable-noinicial> (see Fig. 4):

SADDJ

Fig. 4. Embedded graph for adjectives shortened by apocope inside the SNN

The first node <ADJ+masc+sg> of the following graph is suitable for masculine singular adjectives outside the SNN, e.g., *negro*, in *[[el perro][negro]]* "the black dog" (see Fig. 5):

S

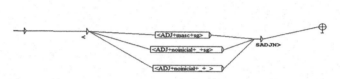

Fig. 5. Embedded graph for masculine singular adjectives outside the SNN

The third node <ADJ+noinicial+_+_> of the following graph is applicable to gender-number invariable adjectives outside the nucleus noun phrase (SNN), like *burdeos*, in *[[los vinos][burdeos]]* "the Bordeaux wines" (see Fig. 6):

S1

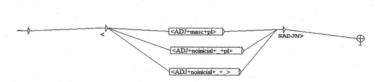

Fig. 6. Embedded graph for masculine plural adjectives, gender invariable adjectives, and number-gender invariable adjectives outside the SNN, the latter nodes with feature [noinicial]

The following graph applies to feminine plural adjectives <ADJ+fem+pl>, e.g. *rosas*, in *[[las sandalias][rosas]]* "the pink sandals," to gender invariable adjectives <ADJ+noinicial+_+pl> , e.g. malvas, in *[[las blusas][malvas]]* "the mauve blouses," and to number-gender invariable adjectives (see Fig. 7):

S3

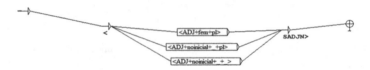

Fig. 7. Embedded graph for feminine plural adjectives outside the SNN

Finally, here is the main grammar (see Fig. 8):

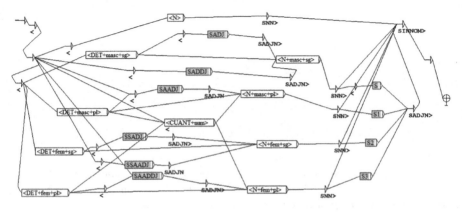

Fig. 8. A syntactic grammar for the noun phrase (SINNOM) in Spanish

Next, it can be clearly observed how the grammar is applied to several phrases, and the analysis it performs (see Figs. 9 and 10):

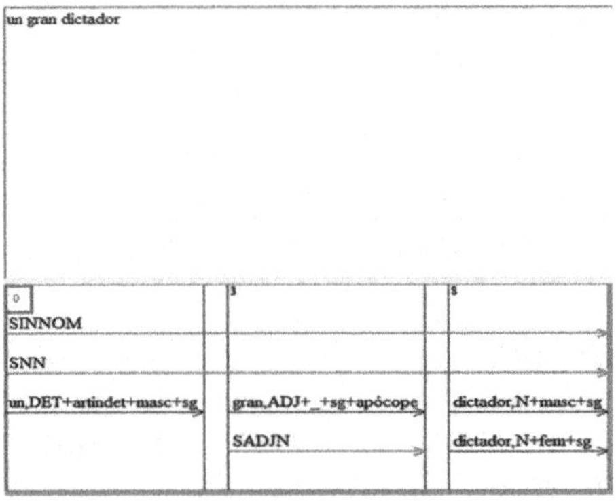

un gran dictador		
0	3	5
SINNOM		
SNN		
un,DET+artindet+masc+sg	gran,ADJ+_+sg+apócope	dictador,N+masc+sg
	SADJN	dictador,N+fem+sg

Fig. 9. Analysis of *un gran dictador* (a great dictator): a nucleus adjective phrase (SADJN) inside a nucleus noun phrase (SNN)

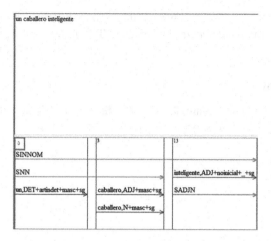

Fig. 10. Analysis of *un caballero inteligente* (an intelligent knight): a nucleus adjective phrase (SADJN) inside the noun phrase (SINNOM)

3   The Adverb in Spanish

Unlike adjectives, adverbs do not present gender and number markings. However, like adjectives, adverbs may have diminutive and superlative degrees. Different types of adverbs are identified and defined according to the combinations they can assume inside the nucleus adverb phrase (SADVN) [2]:

ADV+1: These adverbs always appear alone inside the SADVN, and do not admit modifiers like *equívocamente, finalmente, físicamente*.

ADV+2: These adverbs may either appear alone, or admit modifiers inside the SADVN. Three types of adverbs can be distinguished herein:

ADV+2a (advnuc): they are always the nucleus of the SADVN: *concisamente, abajo, adentro*, but can also be modified by another adverb, like a quantifier: *muy abajo, bien adentro*.

ADV+2b (advmodif): adjectives like *excesivamente* can modify the nucleus, and the nucleus may be an adv2a (advnuc), *abajo*, or an adv2c (adververs), *bien*: *excesivamente abajo, excesivamente bien*.

ADV+2c (ad012), (adververs): adjectives like *completamente* may be the nucleus of the SADVN; but when they are not the nucleus, unlike adv2b, they can either modify an adv2a, *completamente adentro*, or be modified by an adv2b, *casi completamente*.

3.1   The Spanish Adverb in NooJ

The Spanish Module Argentina dictionary of adverbs contains around 400 entries. The distinction between adv1, whose morphological ending is *-mente* (bearing the number 1), and the other adverbs (identified with the number 2) must be included in the dictionary. This distinction is necessary because the adverbs ending in *-mente* cannot appear together consecutively. So, when any combination is proposed, for example, adv2b with adv2a, which is very common, this distinction prevents an ungrammatical combination like *estupendamente divinamente*. Therefore, adverbs are described as follows:

finalmente, ADV + 1
concisamente, ADV + 2a1

Since the morphological grammar used for the adjective is non-specific, it can be modified in order to analyse superlative adverbs (see Fig. 11).

Fig. 11. Analysis of *tempranísimo* (very early), the superlative of the adverb *temprano* (early)

3.2   A Syntactic Grammar for the Nucleus Adverb Phrase

Following [1], the existence of nucleus adverb phrases (SADVN) is assumed. They can either be inside the nucleus verb phrase (SVN), or adjacent to it, inside the verb phrase

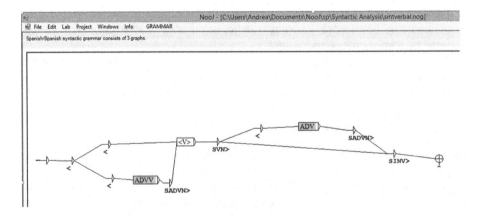

Fig. 12. A grammar for the SINV: the SADVN may be inside SVN, or adjacent to it.

(SINV). Thus, inside the verb phrase (SINV), e.g. *[comen [rápidamente]]*, "they eat quickly," there is a nucleus adverb phrase (SADVN), *[rápidamente]*. It is possible to add a quantifier to the adverb and say: *[comen [muy rápidamente]]*, "they eat very quickly." The quantifier *muy* is used here as a modifier of the nucleus in the nucleus adverb phrase (SADVN).

The main grammar is shown in the figure above (Fig. 12).

After applying Show Debug, two possibilities are observed: adverbs inside the nucleus verb phrase (SVN), and adverbs inside the verb phrase (SINV) (see Fig. 13).

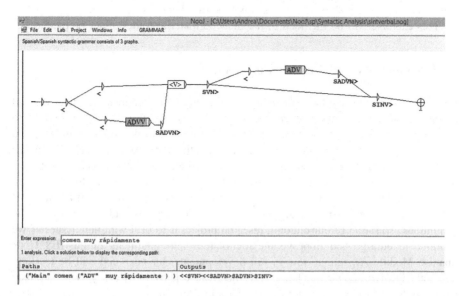

Fig. 13. The expression *comen muy rápidamente* (they eat very quickly) is considered correct.

4   A Pedagogical Application of NooJ

An exercise was proposed for students to apply the grammar to a given text.

4.1   Purposes

– To observe how adjectives that are shortened by apocope appear in real texts.
– To draw conclusions about grammatical constraints in real texts.

4.2   Activity

The text to be analysed is a fragment of *El conde Lucanor: "Lo que sucedió a un hombre bueno con su hijo."* [3] Students were asked to do a search of <ADJ> with Locate. Attention was drawn towards the adjective *bueno* and its particular behaviour. Students had to answer two questions: What happens to the adjective *bueno* in the two occurrences found by Locate? What happens when these occurrences, which are in the singular, are turned into the plural? Then, students were asked to analyse the dictionary entries by pressing CTRL F:

buen, ADJ + masc + sg + apócope
bueno, ADJ + FLX = FLACO + apocopable

4.3 Evaluation

Students were asked to perform a search and look for other adjectives behaving like bueno, and to write a short text using this type of adjectives.

This exercise will be included in the didactic material folder under preparation.

5 Conclusions and Future Perspectives

We have illustrated some elements related to the automatic treatment of adjectives and adverbs in Spanish, by taking advantage of the possibilities offered by NooJ, in order to improve and enhance our research horizon, which is the teaching of Spanish both as a first (L1) and as a second language (L2). An inflectional grammar for the adjective in particular, a morphological grammar that can either be used for the adjective as well as for adverb phrases, a syntactic grammar for the adjective, and another for the adverb were shown. Our work constitutes a first step towards the formalization of the study of adjectives and adverbs, since many aspects have not been yet considered. However, the continuous aspiration of our ongoing research project is to deal with new issues as we continue to enlarge and improve the dictionaries and grammars of the IES_UNR Spanish Module Argentina, now in full development. Our next step will be to study numeral adjectives, on the one hand, and adjectival adverbs, like *llovió fuerte*, "it rained hard," on the other, and to share our progress in future papers.

References

1. Bès, G.G.: La phrase verbal noyau en français. Recherches sur le français parlé **15**, 273–358 (1999). Université de Provence
2. Rodrigo, A.: Tratamiento automático de textos: el sintagma adverbial núcleo. Tesis doctoral. Facultad de Humanidades y Artes. UNR. Ediciones Juglaría, Rosario (2011)
3. Lucanor, C.: Lo que sucedió a un hombre bueno con su hijo. http://www.guillermoromero.es/cuentospopulares/index.php?section=47&page=25

Natural Language Processing Applications

Corpus-Based Extraction and Translation
of Arabic Multi-Words Expressions (MWEs)

Azeddin Rhazi[1(✉)] and Ali Boulaalam[2]

[1] FLSH, Qadi Ayyad University, Marrakech, Morocco
azeddin.rhazi@usmba.ac.ma
[2] FLSF, Med Ben Abdellah University, Fes, Morocco
lingdroit@gmail.com

Abstract. This paper attempts to deal with the problems resulting from the translation of Arabic Multiword Expressions (MWEs), this translation may lead to many difficulties due to the specialized parallel corpus (Arabic-French texts). We first extract monolingual MWEs from each part of the parallel corpus. The second step consists of acquiring bilingual (Arabic-French and Arabic-English) correspondances of MWEs. In order to assess the quality of the mined expression, a statistical and symbolic approach for NooJ Machine Translation (NMT) task-based evaluation is followed. We investigate the performance of a hybrid strategy to integrate extern lexical resources and bilingual MWEs in NMT system. We propose, here, two discriminative strategies to integrate Arabic MWEs in a real parsing context (identification with pre-regrouping and re-ranking parses) with features dedicated to Arabic MWEs. Experimental results show that such a structure as a lexical entry improves the quality of translation.

Keywords: Arabic · MWEs · Corpus · Identification · Extraction
Translation · NooJ

1 Introduction

Multiword Expressions (MWEs) such as "أَخَذَ قِسْطاً مِنَ الرَّاحَةِ *akhada qistan mina ar-rahati*" "he took a break", "رَجَعَ صِفْرَ الْيَدَيْنِ *rajaa sifra al-yadayni*" "he returned empty handed".., often pose problems and difficulties for text analysis chains because of their idiosyncratic nature. Among their notably challenging characteristics, is the fact that they can present irregular syntactic structure, discontinuities and ambiguity. The task of automatic MWEs identification and recognition in the text consists of determining which words are parts of multiword expressions, and how are arguments and predications related to each other.

Considerable progress has been made in the last years to understand and model the interactions between MWEs identification and syntactic parsing. However, the interactions between MWEs and semantic parsing have not received much emphasis for some languages like Arabic. This is surprising, given that many MWEs breach the principle of semantic compositionality to some extent, requiring special treatment in semantic analysis [1].

© Springer International Publishing AG 2018
S. Mbarki et al. (Eds.): NooJ 2017, CCIS 811, pp. 143–155, 2018.
https://doi.org/10.1007/978-3-319-73420-0_12

Therefore, the goal of our contribution is to develop and evaluate original strategies using the NMT for the semantic-aware MWEs identification, parsing and generation of MWEs. Our work follows previous attempts to recognize, extract and translate MWEs using NooJ.

1.1   A Previous Related Research

As described in [2–7], MWEs are treated as a group of simple words that are often used or associated together [5], and are defined as idiosyncratic interpretations that cross word boundaries or spaces [7]. However, it raises some nontrivial issues such as: morpho-syntactic and lexical properties, non-compositionality, idiomacity, syntactic and semantic variation and continuous or discontinuous sequences [8]. The linguistic analysis of various corpora has demonstrated that the great majority of the unknown structures and forms are MWEs which becomes an important problem faced by NLP research on describing languages [4], mainly the morpho-lexical and semantic analysis.

In this paper, we present initial results about parsing the Arabic MWEs using NooJ platform, with respect of their linguistic properties regardless of the statistical aspects, and of the way how NooJ process them. The aim of our study is, also, to clarify up the relative importance of the automatic identification and the extraction of Arabic MWEs from within an online corpus, that is supposed to be analysed and tagged, taking into consideration the multi-word co-occurence and polylexicality and their degree of idiomaticity up to the more frozen in the texts [9].

Our own reaction, in this work, is the implementation of an incremental finite-state transducer-using local grammar of Arabic MWEs [9]. In fact, this work is to improve the Arabic MWEs project that is based on more than 65000 entries. The objective is, on one hand, the identification of a given structures which are classified on a subset of categories using clustering technique; through similar to hierarchical ascending, and on the other hand, the recognition and the formalization of Arabic MWEs external resources [9]. The collected and extracted MWEs from various documents and social network texts are translated in different multilingual purposes.

This article is organized as follows: we describe Arabic MWEs and provide a summary of previous related research. Sub-sections give a brief description of the state of the art and the role of the extern data sources and texts used in this study. Section 2 presents the proposed approach using the NMT. Section 3 exposes some constructed transducers; as preprocessing and processing phases; representing MWEs lexical and translation rules. In Sect. 4 we discuss the results of experiments; preprocessing - regrouping- and processing graphs, and the evaluated relevant results. Finally, we conclude in Sect. 5.

1.2   External Resources and Corpora

The main goal, in this work, is to put the multi-words expressions at the heart of the automatic analysis of texts, and to extract more information typically found in news-paper texts as external resources, from different documents and social network texts. The following objective is the translation of these extracted Arabic MWEs in Mul-tilingual applications environment like Reverso, Google Translate and NooJ (e.g.

extraction and translation), on one hand, and integrating morpho-syntactic and semantic analysis (as a lexical strategy) including NooJ grammatical rules, on the other hand.

In order to achieve our aim, we use NooJ dictionaries and finite state automata only and we rely on a corpus that we have collected from articles that were published in official Moroccan magazines and web pages to be translated by NooJ (Fig. 1) :

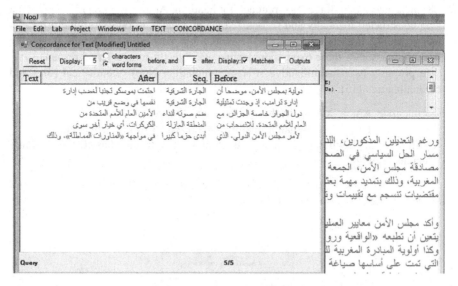

Fig. 1. An extract of Arabic MWEs from Assabah Moroccan Newspaper, May the 2nd, 2017

We have decided to use Arabic texts from this particular magazine, because it is rich with various Arabic MWEs and various forms of expressions, compound words and idioms as shown in Table 1:

Table 1. Presents a sample of Arabic MWEs extracted from the text and translated into French and English

Algérie	Algeria	الجارة الشرقية
Donner une grande importance	Give importance	أبدى حزما كبيرا
Conseil de sécurité	Security Council	مجلس الأمن
L'autonomie locale	Self-government	الحكم الذاتى
Zone isolée	Buffer area	المنطقة العازلة

The corpus has become a good basis for testing the developed NMT system. This evaluation corpus is a collection of parallel texts of MWEs that will be used in later stages of experiments and in future work.

2 Proposed Approach

The proposed methodology adopted both the symbolic and the statistical approaches, which employ information such as parts of speech (POS) filters and lexical alignment between languages. It relies on linguistic methods to identify, and also to use and to produce more targeted candidate lists for extracting Arabic MWEs; both nominal, prepositional and verbal structures; nominal, prepositional and verbal structures [14]. Attia (2008) presented a pure linguistic approach for handling Arabic MWEs [13]; based on a lexicon constructed manually and recognized automatically, in order to build a parallel corpus in Arabic, French and English. We expect this piece of research to highlight the translation function of these expressions online and directly from external resources.

We propose, two discriminative strategies including NooJ grammatical rules and NooJ dictionaries [15]:

– To integrate Arabic MWEs in the real parsing context;
– To process MWEs with semantic features dedicated to Arabic language;
– To immerse MWEs in the semantic analysis and all application based on lexical and semantic strategy.

We hope that this method find solutions to the mentioned difficulties that may result from the automatic translation (see Fig. 2):

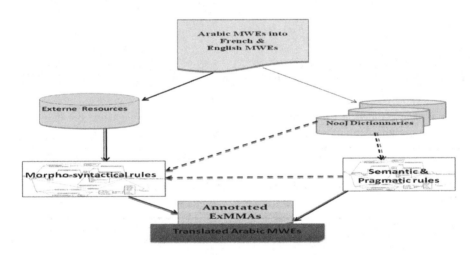

Fig. 2. Processing approach to MWEs translation.

Following the above processing approach, we need to adopt a strategy that consists mainly of manual treatment of a large number of Arabic MWEs as a first step. This methodology demands both symbolic and statistical approach that is based on:

(a) Part of unit information (POS) tagging;
(b) Annotating Arabic MWEs using chunking strategy;
(c) Filters and lexical alignment between languages;

Relying on linguistic methods to produce more candidate lists for extracting Arabic nominal, verbal and prepositional MWEs.

3   Preprocessing and Processing

The NMT that is able to translate MWE structures, is determined by a set of rules developed in the form of syntactic local grammar and morphological local grammar implemented in NooJ system, which states that no grammar rule can be developed independently from a strict delimitation of its domain of implementation [10]. The grammatical and syntactic characteristics for formal and semi-formal modeling of these structures may intervene with both processes of recognition and translation [11].

The model that we propose is used to formalize [8] and to identify Arabic MWEs. We hereby classified Arabic MWEs into six main categories and subcategories [9]:

1- V + N;
2- V + N0 + N1;
3- V + N0 + P;
4- V + N0 + (N + P + N);
5- V + N0 + (P + N + P + P);
6- V + N0 + (N1 + N2),

As the basic classification, we have distinguished several frozen Arabic MWEs, to be extracted and translated, as in the following constructed graphs (presented by Figs. 3, 4, 5, 6, and 7):

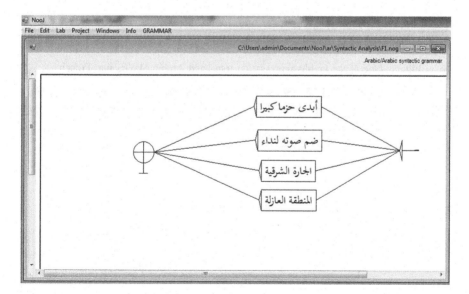

Fig. 3. The graph of four Arabic structures to be translated into French language

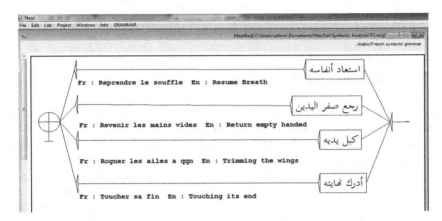

Fig. 4. Graph of French and Arabic MWEs translated into English

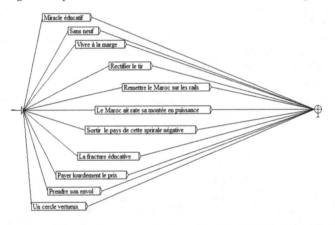

Fig. 5. One of the graphs representing translated Arabic MWEs into French

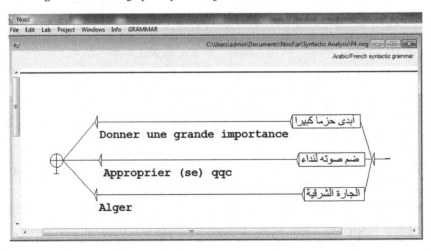

Fig. 6. A sample of translation Arabic MWEs into French

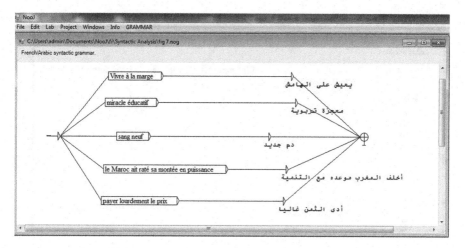

Fig. 7. A second sample of translated Arabic MWEs into French

One of the features of our work is that we used the chunking based semantic segmentation looking to the whole of MWEs in its structural meaning; as a group of word connected together, that can be processed as a single structure that have been already inter-associated as a continuous or discontinuous units [8].

The proposed extraction (Fig. 3) and translations in Figs. 4, 5, 6 and 7 indicate that the semi-automatic recognition of MWE as structures, can simply refer to the appropriate meaning calculated from the whole part of the structure which can be semantically extracted. Similarly, the hypothesis that MWEs can be detected solely by looking at the distinct statistical properties of their individual words, which allows concluding that the association measures can only detect trends and preferences in the co-occurrences of words [1] and can optimize the MWEs translation into a target text.

4 Evaluation and Results

Several problems make the treatment of Arabic MWEs so complicated, and in order to evaluate the accuracy and precision of our hybrid approach, we have created five graphs using NooJ translating rules presented in Sect. 3.

Our evaluation is concerned with the results from the treated examples of the following two experiments setups (Figs. 8, 9):

4.1 Experiment 1

We observe that both Reverso and Google translation of some French MWEs for instance: «brûler l'impatience» and «quitter la vie» ; translated respectively into Arabic language as: «* tahriqu bifaarighi assabri: تحرق بفارغ الصبر» and «* taraka al-hayata: ترك الحياة», adopt the same literal translation, whereas the MWEs does not obey the word-for-word translation, because of their idiosyncratic nature and semantic opacity that produce meaning from all the components and the whole part of the expressions.

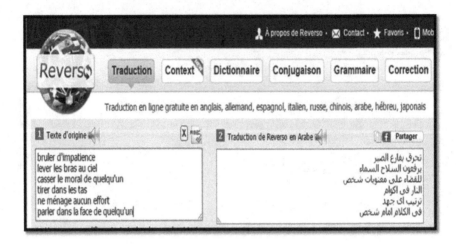

Fig. 8. Reverso translation experiment of some Arabic MWEs

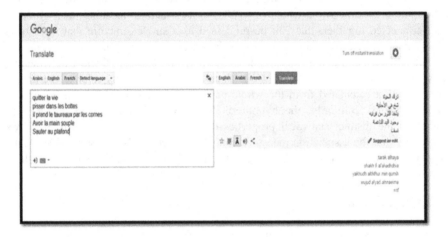

Fig. 9. Google translate experiment of some Arabic MWEs

4.2   Experiment 2

In contrast to the above experiment, the NMT system optimizes the translation of Arabic MWEs as in the following results (see Figs. 10, 11, 12, 13, and 14):

In this sense, and in view of the specificities of fixed expressions, a hybrid approach has been adopted to combine the linguistic aspects and the statistical approach. Consequently, for the extraction task as for the translation task, we have used the NooJ platform. This approach is based on two main mechanisms, namely:

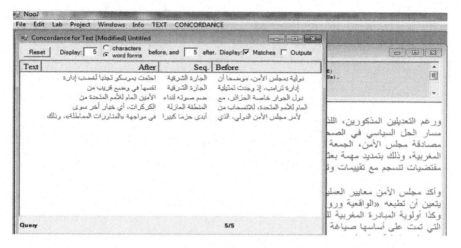

Fig. 10. Arabic MWEs concordance

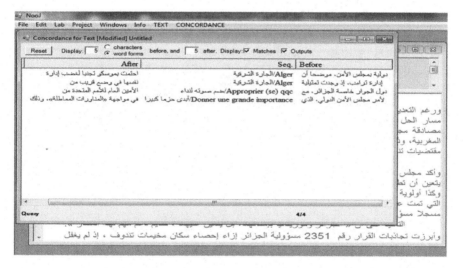

Fig. 11. Translated MWEs into French

- The development of a comprehensive database covering the different categories of fixed expressions, based on the concept of the "fixed area" [12] recognized automatically from the whole of the structure;
- The integration of changes and potential transformations of fixed expressions [14] to widen the margin of choice of suitable equivalences within the target language.

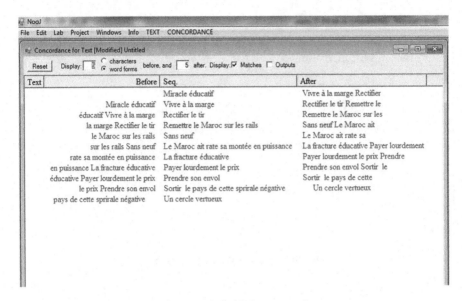

Fig. 12. MWEs translation concordances

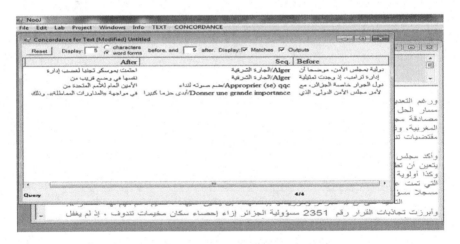

Fig. 13. Translation results

This approach, also, has enabled us to achieve the following results:

(a) The translation for several equivalencies produced by the two systems (see experiment.1) shows that:

1. The majority of the expressions produced in the target language are incorrect;
2. Some of the expressions are translated literally;
3. The absence of a mechanism for the arrangement of parts of words according to the specific characteristics of the target language.

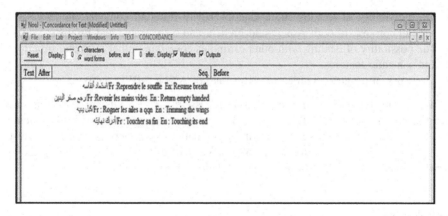

Fig. 14. Results of the parallel MWEs translation

(b) While we could experience positive developments in this translation (from Arabic to French or English and vice versa) using NMT rules and dictionaries we observe that:

1. The Automatic translation of multiword expressions requires a combination of statistical methods and linguistic methods. This hybrid approach contributes to the removal of ambiguity after a lexical analysis via external resources.

2. The use of a segmentation based on the chunking instead of the morpho-syntactic analysis, which consists of splitting the structure into non-recursive simple constituents called chunks. This operation has allowed us to determine the main element of expression as a trigger during translation; associating semantically the arguments to the predicates.

3. Experimentation; with some software such as Reverso and Google; shows that they do not take into account the characteristics of non-compositionality of the MWEs which does not obey the word-for-word translation.

4. This has led us to adopt an approach based on the search for equivalences between source and target languages as we do using the NMT system.

5 Conclusion and Perspectives

In this paper, we presented the difficulties of Arabic translating of MWEs. We have concluded that using the NMT mainly its translation rules as morpho-syntactic and semantically based rules [8], enable us to translate Arabic MWEs more accurately by relying on:

- The development of a system that allows the translation of MWEs Arabic into French and English languages.
- The potentiality of NooJ Machine Translation that is flexible, versatile, and easy to use and to optimize online.

- The possibility of treating more examples with different contexts and large and parallel corpora, in both Arabic and French, and also Arabic and English.
- The possibility of dealing with other examples containing other syntactic structures and semantic features.
- The work in progress with a large coverage of Arabic MWEs corpora.

We aspire to highlight the translation process of Arabic texts by extracting and recognizing Arabic MWEs, from parallel and extern resources directly, by expanding the NooJ dictionaries with new entries in a systematic way (covering large and diverse areas of the lexicon's inventory of MWEs). We realize that NooJ experiment shows parsing efficiency in the extraction and translation of Arabic MWEs.

References

1. Meghawri, S., et al.: Semantic extraction of Arabic multiword expressions. Computer science and Information Technology (CS&IT), pp. 23–31 (2015). https://doi.org/10.5121/csit.2015.50203
2. Najar, D., Mesfar, S., Ghezela, H.B.: A large terminological dictionary of Arabic compound words. In: Okrut, T., Hetsevich, Y., Silberztein, M., Stanislavenka, H. (eds.) NooJ 2015. CCIS, vol. 607, pp. 16–28. Springer, Cham (2016). https://doi.org/10.1007/978-3-319-42471-2_2
3. Constant, M.: Mettre les expressions multi-mots au cœur de l'analyse automatique de textes: sur l'exploitation de ressources symboliques externes. Université Paris-Est (2012)
4. Attia, M., et al.: Automatic extraction of Arabic multiword expressions. In: Proceedings of the Workshop on Multiword Expressions: from Theory to Applications, Beijing. August-MWE, pp. 18–56 (2010)
5. Pecina, P.: Lexical association measure: collocation extraction. Institute of Formal and Applied Linguistics (2009). Editor in chief, Jan Hajič
6. Baldwin, T.: Multiword expressions, an advanced course. the Australasian language technology summer school (ALTSS). Sydney, Australia (2004)
7. Sag, I.A., Baldwin, T., Bond, F., Copestake, A., Flickinger, D.: Multiword expressions: a pain in the neck for NLP. In: Gelbukh, A. (ed.) CICLing 2002. LNCS, vol. 2276, pp. 1–15. Springer, Heidelberg (2002). https://doi.org/10.1007/3-540-45715-1_1
8. Silberztein, M.: La formalisation des langues: l'approche de NooJ. Collection sciences cognitive et management des connaissances. Edition ISTE, London (2015)
9. Rhazi, A.: Morpho-syntactical based recognition of Arabic MWEs using NooJ platform. Formalizing natural languages In: Proceedings of the 2014 International NooJ Conference, Cambridge Scholars Publishing, Newcastle (2015)
10. Gross, M.: The construction of local grammars. In: Roche, E., Schabs, Y. (eds.) Finite State Language Processing, pp. 329–354. MIT Press, Cambridge (1997)
11. Fehri, H., et al.: A new representation model for the automatic recognition and translation of Arabic named entities with NooJ. In: Proceedings of the 9th International Workshop on Finite State Methods and Natural Language Processing, ACL, Blois, France, 12–15 July, pp. 134–142 (2011)
12. Laporte, E.: La reconnaissance des expressions figées lors de l'analyse automatique. Langage, no. 90 (1988)

13. Attia, M., et al.: Handling Arabic Morphological and Syntactic Ambiguity within the LFG Framework with a View to Machine Translation. Ph.D. thesis. The University of Manchester, Manchester, UK (2008)
14. Kocijan, K., Librenjak, S.: Recognizing verb-based Croatian idiomatic MWUs. In: Okrut, T., Hetsevich, Y., Silberztein, M., Stanislavenka, H. (eds.) NooJ 2015. CCIS, vol. 607, pp. 96–106. Springer, Cham (2016). https://doi.org/10.1007/978-3-319-42471-2_9
15. Silberztein, M.: La formalisation du dictionnaire LVF avec NooJ et ses applications pour l'analyse automatique de corpus dans Langages. Numéro 2010/3, pp. 179–180 (2010)

The Automatic Translation of French Verbal Tenses to Arabic Using the Platform NooJ

Hajer Cheikhrouhou[(✉)]

University of Sfax, LLTA, Sfax, Tunisia
cheihkrouhou.hager@gmail.com

Abstract. Our ultimate purpose in this work is the study of the French temporal system in parallel with the Arabic one. This parallelism revealed similarities as well as differences concerning the functioning of both verbal systems at the level of the transformations in the verbal sentence made by mood, tenses and aspects. In this article, we will try to clarify the similarities and the differences between both temporal systems and the corresponding tenses in Arabic. Second, we are going to create a machine translation system where we try to consider the temporal specificities of every language and to solve the problem related to the agreement in person and number.

Keywords: French verbal tenses · Arabic verbal tenses · Automatic translation NooJ

1 Introduction

The creation of machines capable of suitably translating the French verbal tenses is based on two steps: a theoretical step and an applied one. The first phase enables us to analyze and interpret the linguistic phenomena and the second phase is devoted to the realization of the automatic translation system.

Our work is situated within the framework of applied linguistics and we seek:

- The creation of a system of analysis and recognition of syntactic constructions of French verbs. As a case study, we have chosen the class of verbs of communication (C) and movement (E) according to the classification of "French verbs" of Dubois and Dubois-Charlier [1] representing an exhaustive syntactic-semantic classification of verbs.
- The automatic recognition of verbal tenses.
- An adequate and viable automatic translation of verbs using the platform NooJ.

2 The Problems of Automatic Translation

The translation from a source language to a target language and essentially the translation of verbs faces many difficulties created by the ambiguity of the source language and essentially by the differences between both languages.

S. Mbarki et al. (Eds.): NooJ 2017, CCIS 811, pp. 156–167, 2018.
https://doi.org/10.1007/978-3-319-73420-0_13

2.1   Lexical Divergence

Example: "*décaserner*" means "Ôter de la caserne»: to remove someone from the barracks.

– The locative is integrated in the semantics of the verb
– The prefix "*dé*" expresses the action of removing someone.

In Arabic, this structure is impossible:
Décaserner = أَخْرَجَ مِنَ الثُّكْنَة (faire sortir de la caserne = to remove from the barracks)
Example: Il décaserne les soldats. أَخْرَجَ الجُنُودَ مِنَ الثُّكْنَةِ
(He removes soldiers from the barracks).

2.2   Temporal Divergence

Example: *J'avais discuté* cette idée avec le directeur.

In Arabic, the past perfect (le plus-que parfait) is translated by [2]:
«*kāna qad fa'ala* : كَانَ قَدْ فَعَلَ» + the value of anteriority

Example: *kuntu qad nāqaštu* hadihi l-fikrata ma'a l-mudīri.

كُنْتُ قَدْ نَاقَشْتُ هَذِهِ الفِكْرَةَ مَعَ المُدِيرِ = I had discussed this idea with the director.

- *Kāna* is used to show anteriority.
- *Qad* expresses modality and aspect.
- *Fa 'ala* marks the finished aspect.

Example: Le professeur *explique* cette théorie aux étudiants.
The professor explains this theory to the students.

The present corresponds to « *muḍāri'* marfū': yaf'alu » (nominatif inaccompli)
الأستاذُ يُفَسِّرُ هذه النظرّية إلى الطلبةِ.
'al-'ustāḏu *yufassiru* haḏihi l-naḍarriyata'ilā ṭalabati.
The action yaf'alu: يَفْعَل (yufassiru = explains) can indicate two times:

– *The present:* 'al-ḥāḍir
– *The Future:* 'al-mustaqbal

To state precisely the verbal tense, we use temporal expressions such as '*al- yawma* (aujourd'hui = today), *ġadan* (demain = tomorrow),'*al-'āna* (maintenant = now)...
Example:
الأستاذُ يُفَسِّرُ هذه النظرّية إلى الطلبةِ اليَوْمَ.
'al-'ustāḏu *yufassiru* haḏihi l-naḍarriyata'ilā ṭalabati'al- yawma.
Today, the professor explains this theory to the students.

⟶ The action is taking place in the present.

الأستاذُ يُفَسِّرُ هذه النظرّية لِلطَّلَبَةِ غَدًا.

'al-'ustāḏu yufassiru haḏihi l-naḏarriyata'ilā ṭalabati ġadan.

Tomorrow, the professor will explain this theory to the students.

⟶ The action will take place in the future.

Or the context and the enunciative situation state precisely time variables.

3   A Comparison Between the French and the Arabic Verbal Systems

The verb is a grammatical category which plays a fundamental role in the sentence. It is the syntactic and semantic nucleus of the proposition to which it gives a pragmatic anchorage by the morphological marks of person, number, tense, mood and voice.

One of the principal functions of the conjugated verb is to situate the action in time. Indeed, the verb gives indications about the realities it designates which are of three types: mood, tense and aspect.

3.1   Mood

As it is traditionally known, the French verbal system is divided in personal moods (Indicatif, Subjonctif et Impératif) and each mood comprises its proper times where the aspect is accomplished (passé composé, plus que parfait) or unfinished (present, future...).

According to the traditional Arabic grammar, the moods, called «ṣ-ṣiyaġu: الصّيغ» are: *The past* (māḏī), the present (muḏari') and the order ('amr) which are in close relation with the aspect of the verb: finished (complete) or unfinished where the actions are respectively: «fa'ala: فَعَل» and «yaf'alu : يَفْعَل» [3].

The Accomplished Aspect. This aspect is linked to a single mood which is the past (le passé), it is le *māḏī*.

Example: ذَهَبَ إلَى السِّينِمَا ḏahaba 'ilā s-sinīmā.
           Il est allé au cinéma = He went to the cinema.

The Unfinished Aspect. It is related to two moods: *'al-muḏāri'u* (المُضَارِعُ) and l'impératif: *'al-'amru* (الأَمْرُ).

The *muḏāri'u* has three forms: marfū'(مرفوع), manṣūb (منصوب) et maǧzūm (مجزوم) respectively called: inaccompli nominative, inaccompli apocopé and inaccompli accusatif.

Example: *yaktubu ris*ālatan liṣadīqihi. يكتبُ رسالة لصديقه
           Il écrit une lettre à son ami. He is writing a letter to his friend.
Example: *'uktub° ris*ālatan liṣadīqika. أُكْتُبْ رسالة لصديقك Impératif (*'amr*)
           Ecris une lettre à ton ami.
           Write a letter to your friend.

3.2 Tense

In French, time is determined by:

- The simple form where the aspect is unfinished.
- The compound forms where the aspect is accomplished.

Within the same vein of thought, in his classification of tense in Arabic, in his book «šarḥu l-mufaṣṣali», Ibn Yaîch indicates that there are three tenses: the past, the present and the future.

The past: (māḍin): *faʿala*. Example: ḍahaba → ذَهَبَ (Il est allé.) = He went.

The present: (ḥāḍirun): *yafʿalu*. Example: yaḍhabu → يَذْهَبُ (Il va.) = He goes.

The future: (mustaqbalun): Sa/Sawfa yafʿalu.

Example: **Sayaḍhabu** → سَيَذْهَبُ (Il ira) / *Sawfa* yaḍhabu → سَوْفَ يَذْهَبُ = He will go.

As for the conjugation of the compound forms, the French verbal system uses the auxiliaries "avoir" and "être" to express the temporal varieties of the past tense. In Arabic, we use the verbal form «*kāna* : كَانَ», the modal particle «*qad* : قَد» or the two at the same time «*kāna qad* : كَانَ قَد» to express the temporal value of anteriority.

- The verbal form «*kāna* : كَانَ» anchors the action in the past.
- The modal particle «*qad* : قَد» carries the aspect and the mood.

In the past, «*qad* : قَد» confirms the complete aspect of the action (completed action) and the actual. With the present, it reinforces the unfinished aspect of the verb and it turns the action probable or virtual.

Example: قَدْ نَجَحَ أَحْمَدُ فِي الأُمْتِحَانِ.

qad naǧaḥa'aḥmadu fī l-'imtiḥāni. (Ahmed a réussi à l'examen.)
Ahmed passed the exam.

قَدْ يَنْجَحُ أَحْمَدُ فِي الأُمْتِحَان.

qad yanǧaḥu'aḥmadu fī l-'imtiḥāni. (Ahmed réussira (peut être) à l'examen.)
Ahmed would pass the exam.

In fact, every translation operation shows similarities but especially differences between the two languages dealt with such as marks of agreement in person and number and also negation which presents a great difference between the two verbal systems.

3.3 Agreement

In French, verbs have the same morphology with masculine or feminine subjects with the exception of the past participle (participle passé) conjugated with the auxiliary "être"

In Arabic, the verbs are not conjugated in the same verbal form with masculine or feminine subjects (see Fig. 1).

We remark that in Arabic the marks of agreement in person and number are attached to the verb and are explicit even if the subject is not stated. On the other hand, in French, verb endings do not always change in person which engenders another ambiguity in the creation of an automatic translation system.

Il / elle discute	He discusses/She discusses		Tu discutes	You discuss
Masculine	huwa yunāqišu	هُوَ يُنَاقِشُ	'anta tunāqišu	أَنْتَ تُنَاقِشُ
Feminine	hiya tunāqišu	هِيَ تُنَاقِشُ	'anti tunāqišīna	أَنْتِ تُنَاقِشِينَ

Vous discutez	You discuss		Ils/Elles discutent	They discuss
Masculine	'antum tunāqišūna	أَنْتُم تُنَاقِشُونَ	hum yunāqišūna	هُم يُنَاقِشُونَ
Feminine	'antunna tunāqišna	أَنْتُنّ تُنَاقِشْنَ	hunna yunāqišna	هُنّ يُنَاقِشْنَ

Fig. 1. The agreement marks in person and number in Arabic.

3.4 Negation

One last point which highlights the difference between the two verbal systems is that of negation in the two languages. In French, negative particles « ne –pas/plus/jamais…» are valid for all tense variables:

Example: Il n'a pas dit. Il ne dit pas. Il ne dira pas.
         He did not say. He does not say. He will not say.

On the other hand, negation in Arabic marks changes with the tense of the verb:
Verb in the present: « lā : لا » ou « mā : ما »+ *yaf'alu*
            inaccompli nominatif: muḍāri' marfū'

Example: لا أُنَاقِشُ هَذِهِ الفِكْرَةَ مَعَ المُدِيرِ lā'unāqišu ḥadiḥī lfikrata ma'a lmudīri.
         Je ne discute pas cette idée avec le directeur.
         I do not discuss this idea with the director.

Verb in the past: « lam : لَم » + *yaf'al°*
            inaccompli apocopé: muḍāri' maġzūm

Example: لَم أُنَاقِشْ هَذِهِ الفِكْرَةَ مَعَ المُدِيرِ Lam'unāqiš° ḥadiḥī lfikrata ma'a lmudīri.
         Je n'ai pas discuté cette idée avec le directeur.
         I did not discuss this idea with the director.

Verb in the future: « lan : لَن » + *yaf'ala inaccompli accusatif: muḍāri' manṣūb*

Example: لَن أُنَاقِشَ هَذِهِ الفِكْرَةَ مَعَ المُدِيرِ Lan'unāqiša ḥadiḥī lfikrata ma'a lmudīri.
         Je ne discuterai pas cette idée avec le directeur.
         I will not discuss this idea with the director.

At the end of this comparative study, we can say that we are dealing with two different verbal systems in terms of tense, aspect and mood. This difference goes back to the Indo-European origin of the French language and the Semitic origin of the Arabic one. But there is no denying that there are similarities between both languages already identified in this section.

In what follows, we are going to create an automatic translation system, using the platform NooJ, where we attempt to take into consideration the temporal specificities of every language and to solve the problems related to agreement in person and number, according to the results obtained from our first comparative study of the two verbal systems.

4   The Process of French-Arabic Automatic Translation

Realising an application of automatic translation with NooJ consists of constructing the linguistic data:

- An electronic bilingual dictionary
- A bilingual formal grammar

Our program of automatic translation is based mainly on the implementation of verbs of communication [4] and of movement [5] in two bilingual dictionaries French-Arabic which serve first to reformulate the information provided in LVF by the operators and the language of NooJ. Then we will create formal grammars to remove any syntactic ambiguity and finally we will automatically translate these predicates into Arabic (see Fig. 2).

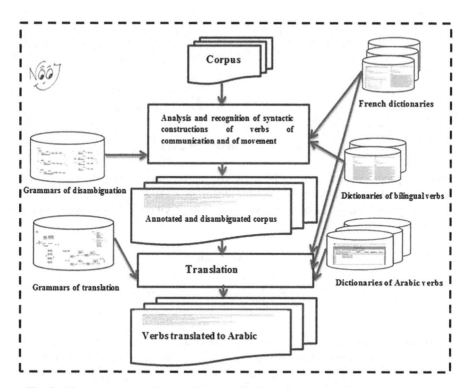

Fig. 2. The proposed architecture for the realization of an automatic translation system.

4.1   The Creation of Grammars for the Analysis and Recognition of Syntactic Patterns

In this phase, we will try to create formal grammars able to analyse the sentences. This aims at recognizing the different arguments found in a corpus or a text and sorting the suitable syntactic constructions without any ambiguity.

The constructed grammars are called grammars of syntactic disambiguation [6] (see Fig. 3).

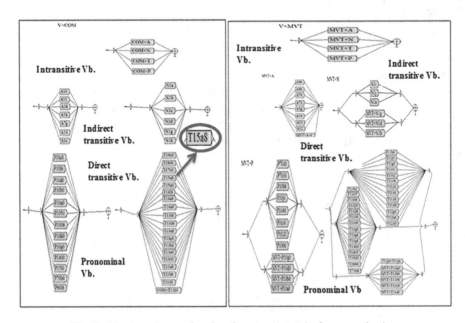

Fig. 3. The transduces of verbs of movement and of communication.

As a syntactic construction, we have created the argument structure [T15a8]. The verbs described by this construction subcategorize certain complementations like a complement with «que + subjunctive» or an infinitive with «de» with a prepositional phrase «à qn: to someone».

Example: Le professeur demande à l'élève le silence
            Le professeur demande à l'élève de faire silence.
            Le professeur demande à l'élève qu'il fasse silence.
            The professor asks the pupil to be quiet/to keep quiet.

To remove ambiguities, we make use of a grammar of disambiguation which strains annotations and keeps only those belonging to the construction [T15a8] (see Fig. 4).

In this grammar, we identified the semantic information of the prepositional phrase (indirect object) by [N + Hum] introduced by the preposition "à". As for the direct object, we described the three possibilities: infinitive proposition, nominal group or completive proposition.

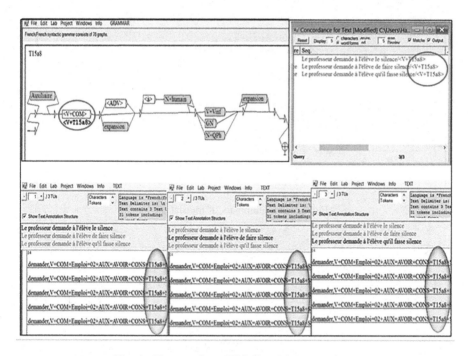

Fig. 4. The transduce [T15a8] and its annotations.

The displayed annotations show that the sentences are analyzed only by the right argument structure [T15a8].

This first step of analysis and automatic recognition of different argument structures is very important in the creation of the process of automatic translation because the more the annotations are precise and right, the more the results of the automatic translation in the second phase are adequate. Indeed, the results of the first phase are the entries of the second step.

4.2 The Creation of Grammars of Automatic Translation

This step is based on the same lexical resources but with the addition of dictionaries of Arabic verbs which will provide conjugated forms according to the indications provided by the grammars of translation. To accomplish this step, we have created transduces of translation of verbs in both forms, affirmative and negative as there are some differences between the two systems at this level (see Fig. 5).

The Translation of Verbs in the Affirmative Form. The translation of a verb depends on the subject in terms of person and number. This condition obliges us to construct grammars of translation which take into consideration the subject which can be a human, a personal pronoun or a proper noun. For this reason, we put <PRO + nom>,<N + hum> et <N + PRENOM> in variable <S> to indicate the nature of the subject (see Fig. 6).

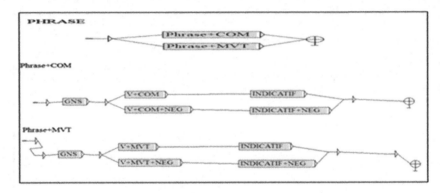

Fig. 5. The grammars of translation.

GNS

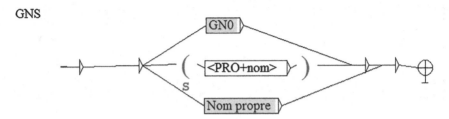

Fig. 6. The variable < S>.

Before verb translation, the system must go through two steps:

<*S Genre = m* > and < *S Genre = f*>.

This operation consists in deducing whether the subject is feminine or masculine.

After this step, the system will do tests to verify the person and tense of the verb which is put in variable < V >.

Example: < *$V:V + PR + 3+s* > where the variable < V > is a verb in the simple present with the third singular personal pronoun.

If this test is validated, the system moves to the third step of translation:

<*VAR_V + A+P + 3+s + m* > which means translating V to AR (Arabic) in the active form, in the present with the third singular masculine personal pronoun (see Fig. 7).

The application of these grammars on sentences, where the verb is conjugated in the simple present, provided the following results (see Fig. 8):

Présent de l'indicatif

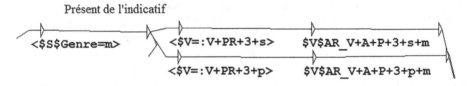

Fig. 7. The transduces of translation of the simple present in the affirmative form.

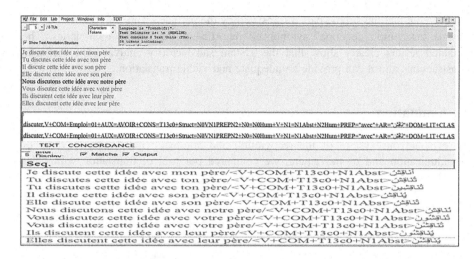

Fig. 8. The translation of « discuter + T13c0 » . (to discuss)

We found forms conjugated with all the personal pronouns with varieties in person.

The Translation of Verbs in the Negative Form. We remember that the negation in Arabic is different from that in French. Indeed, the negation in Arabic is in connection with the tense of the verb where we use the negative particules « lā, lam et lan : لَا - لَنْ – لَمْ ».

For example, the negative particle « lam: لَمْ » is used to negate actions taking place in the past (see Fig. 9).

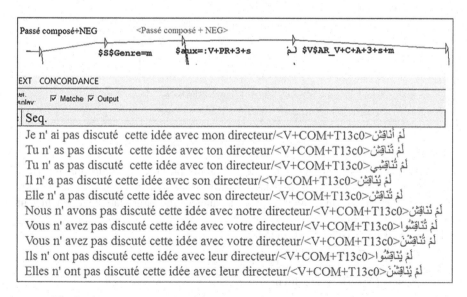

Fig. 9. The transduces of translation of the simple past in the negative form.

We note that the simple past is translated by the particle « lam: لَمْ» followed by the "forme inaccompli apocopé" symbolised by <C> in the Arabic dictionaries. The obtained results clearly show that the system satisfactorily analyzed the syntactic constructions and it can provide an adequate and viable translation.

5 Evaluation

To evaluate our system of the recognition and automatic translation of verbs of communication and of movement, we compared our tool of translation with other transductors like Google and Reverso (see Fig. 10).

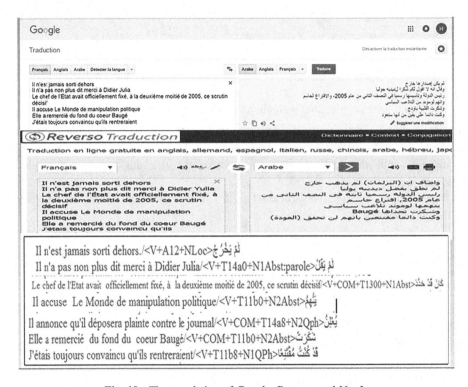

Fig. 10. The translation of Google, Reverso and NooJ.

We notice that the translation by these systems provides aberrant and incorrect results contrary to our system of translation NooJ. In fact, we notice that these systems did not provide translations which respect the rules of the target language whether in terms of tense or meaning.

Example: Il n'est jamais sorti dehors

→ لم يكن اصدارها خارج (google)

→ و أضاف أن (البرلمان) لم يذهب خارج (Reverso)

→ لَمْ يَخْرُجْ (NooJ) = He did not go

Example: Le chef de l'Etat avait officiellement fixé, à la deuxième moitié de 2005, ce scrutin décisif.

(google) رئيس الدولة و تأسيسها رسميا في النصف الثاني من عام 2005   →

(Reverso)   رئيس الدولة رسميا ثابتة في النصف الثاني من عام 2005 , اقتراع حاسم   →

(NooJ) = He had fixed   كانَ قَدْ حَدَّدَ   →

We can affirm that the system of translation elaborated, using the platform NooJ, clearly shows its efficiency in the phase of analysis and recognition of different syntactic constructions and also in producing correct and adequate translations which respect the specificities of the target language.

6   Conclusion

In this work, we have tried to create an automatic translation system NooJ based on two phases: a phase devoted for the analysis and the recognition of different syntactic constructions and a phase of translation.

The implementation of the automatic translation process enables us to find encouraging results where the verbs of communication and of movement are translated in Arabic with respect to the rules of conjugation of the target language in its affirmative and negative forms.

Evidently, much has to be done. For the rest of this work, we think of automatically translating other verbal classes to Arabic and the creation of other bilingual French-Arabic dictionaries.

References

1. Dubois, J., Dubois-Charlier, F.: Les verbes français. Larousse, Paris (1997)
2. Cheikhrouhou, H.: Arabic translation of the french auxiliary: using the platform NooJ. In: Barone, L., Monteleone, M., Silberztein, M. (eds.) NooJ 2016. CCIS, vol. 667, pp. 74–86. Springer, Cham (2016). https://doi.org/10.1007/978-3-319-55002-2_7
3. Chairet, M.: Fonctionnement du système verbal en arabe et en français (Linguistique constractive et traduction). Éditions OPHRYS_Paris, France (1996)
4. Cheikhrouhou, H.: Recognition of communication verbs with NooJ platform. In: Formalising Natural Languages with NooJ 2013. Cambridge Scholars Publishing, British, pp. 155–169 (2014)
5. Cheikhrouhou, H.: The formalisation of movement verbs for automatic translation using NooJ platform. In: Formalising Natural Languages NooJ 2014. Cambridge Scholars Publishing, British, pp. 14–21 (2015)
6. Silberztein, M.: La formalisation des langues l'approche de NooJ. In: Collection Science Cognitive et Management Des Connaissances. ISTE Editions (2015)

Automatic Extraction of the Phraseology Through NooJ

Tong Yang[⊠]

DILTEC (Didactiques des langues, des textes et des cultures),
Université Sorbonne Nouvelle (Paris 3), Paris, France
tongyangparis3@gmail.com

Abstract. To teach the nominal expressions of the phraseology to foreign learners, we must build a corpus from which we can extract desirable sequences. Modeling and disambiguation are at the heart of extraction. In this article, we discuss how the two procedures are established and also show how a data implementation is processed in *NooJ*. At the end of the article, our quantitative and qualitative analyses prove that the result of this extraction is positive.

Keywords: Cuisitext · Phraseology · Extraction · Modeling · Disambiguation

1 Introduction

Our study is part of the teaching of the FOS (French for specific purposes, 2004) to foreign cooks coming to work in French restaurants. Due to fact that the number of polylexical units is very large in comparison to that of monolexical units in the same part of the discourse, the existence of polylexical sequences has been taken into account since the end of the 20th century [2, 10, 13, 14, 17, 20, 21]. [22] had been the first one to gather all non-free expressions under the label "phraseology" while many scholars came later to stress the importance of teaching phraseology for specific purposes [2, 3, 11]. In this sense, in the teaching of polylexical expressions in the culinary domain, the extraction of the expressions of the discourse has become an unavoidable task. Nonetheless, the extraction of multi-word sequences still poses a major problem [19]. The purpose of this article is to reach the best method towards an exhaustive extraction of the nominal expressions of the phraseology in the culinary domain using a tool of TAL (automatic language processing).

2 Study Corpus

Our corpus named *Cuisitext* [29] contains both written and oral corpuses. For the written part of the corpus, which we have collected, using the creator of corpus *gromotor*, was collected from thousands of French recipes from French culinary sites, such as "*Marmiton*", "*750 g*", "*Cuisine AZ*", etc. In the professional setting of cooks,

© Springer International Publishing AG 2018
S. Mbarki et al. (Eds.): NooJ 2017, CCIS 811, pp. 168–178, 2018.
https://doi.org/10.1007/978-3-319-73420-0_14

communication is usually oral. Thus, our corpus has three types of oral corpus in video format and we present them in Table 1 below.

Table 1. Three types of oral corpus in our study

Oral corpus	Number of videos	Mins/ video	Year	Source
Culinary videos on the Internet	1 Hundred	5	After 2010	TV channels
Films in a hotelschool	Five	1	2016	Mangiante J.M. (University of Artois)
Films in two kitchens	Twenty	60	2015	Ourselves

The culinary video clips were selected from the Internet according to the three types of French dishes presented previously. The second type of videos was filmed in the classes of a hotelschool offered by Mangiante. In addition to that, we have 20 h of recording, which we have carried out, in two kitchens, during the opening hour of two restaurants in Font Romeu at the end of December 2015. The elaboration of an oral corpus imposes a transcription for the didactic exploitation. Thanks to the software CLAN, the transcription of our oral corpora was carried out correctly. Once the constitution of *Cuisitext* is completed, we are struck by the high frequency of nominal expressions (e.g., *steak haché, ail bienhaché, cake beurré*). Before linguistic analysis and didactic aspect are taken into consideration, the first question to be asked is that of the selection of the appropriate extraction software.

3 The Selection of Extraction Software

3.1 Key of Extraction: Semi-phraseological Expressions

These expressions are part of the expressions of phraseology, but the definition of phraseology is broad and vague, so a linguistic description of nominal expressions is required. A linguistic treatment allows us to take into account the syntactic and semantic relations between the names and adjectives of the expressions in order to be able to grasp the key of the extraction. Indeed, the question of typology in phraseology has been dealt with extensively by several linguists, e.g., [8, 10, 21, 28], specially [23]. He divides the expressions of phraseology into two main categories: phraseological expression and semi-phraseological expression. The first is synonymous with the idiomatic expression, which is often used in linguistic literature. The relations between the elements of an idiomatic expression are fixed at the syntactic and semantic level.

However, in a semi-phraseological expression, only one of the elements must be chosen in a constrained way according to other elements of this expression. In this sense, collocation is part of the semi-phraseological expression. Our extracted expressions thus contain both phraseological expressions (e.g., steak haché) and semi-phraseological expressions (e.g., ail bienhaché, cake beurré). At the syntactic level, phraseological expressions do not accept any external element between their components, but the semi-phraseological expression are not constrained and their elements are not necessarily contiguous. As authors [16] say, the semi-phraseological expression is the central problem of fixed expression. According to [23], the semi-phraseological expression is, in a sense, a "à trous" expression that functions as a linguistic sign, but the distance between elements is not measurable and this non measurable distance constitutes the difficulty of the extraction. The central problem of extracting phraseology is of semi-phraseological expressions, and then the search for the appropriate software becomes a challenge to this extraction.

3.2   Functions and Programs Developed for Extraction

Because of the lack of computer knowledge, didacticians, terminologists and translators use functions developed for extracting polylexical sequences, such as "n-gram" of [1] or "repeated segments" of *Lexico 3* [25]. The two functions can detect the lexical combinatories that remain frequent in a corpus textual space, but the result is often very frustrating and unusable because, on the one hand, this detected text space is limited to 5 words. The problem then is that idiosyncratic interpretations exceed this limit of words [9, 24]. On the other hand, grammatical words can intervene in the extraction so that false semi-phraseological sequences can be output, such as "de la", "dans un", because in these treatments the sequences are always considered as the set of letters and spaces. Some specialists of TAL have become aware of this problem and devoted their time to the development of a software for the extraction of polylexical expressions [9, 15, 18, 27].

The development of this software had to follow two main procedures. Firstly, only statistical analyses (for example, *Z-score*, *Mutual Information*) were integrated for this extraction [4, 5, 7]. There is no doubt that this integration ensures the accuracy of the result of the extraction, but it still lacks linguistic analyses, which every linguist is regularly confronted with. Secondly, specialists associated linguistic treatment with statistical analysis in developed programs, for example, *programof Lin* [18], *FipsCo* [9], *programof* [12]. The first one can analyze sentences of more than 25 occurrence of words and the second can perform a number of grammatical transformations for the treatment of the direct subject. We emphasize that the transformations of a sentence and the disambiguation of the names and adjectives are the most fundamental tasks in our extraction. However, we did not find a flexible software that can accomplish both tasks in an infinite space of a corpus. Thus, we have discovered a language training software capable of describing languages exhaustively: *NooJ* [26].

3.3   Approach of NooJ

As a corpus processing system, *NooJ* can make the disambiguation and transformation of sentences, and it can offer corpus processing possibilities for teaching [26]. Based on the Chomsky-Schützenberger hierarchy [6], *NooJ* described all the generative grammars: unrestricted grammar, contextual grammar, algebraic grammar and rational grammar (in Fig. 1 below).

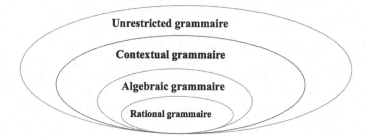

Fig. 1. Four types of generative grammars

The four types of generative grammars have a hierarchical relation: the set of rational grammars is included in the set of algebraic grammars, which itself is included in the set of contextual grammars and the set of unrestricted grammars. In theory, *NooJ* can describe all the languages in the world with these four types of generative grammars [26] and in *NooJ*, grammar can be expressed in two forms: textual grammar and graph grammar.

In addition to that, the results of phrase extraction must be evaluated at two levels: the exhaustiveness of the extraction and the success rate of the result. Neither can be ignored. Inspired by [28], we aim to make a fine modeling of expressions of phraseology so that the extraction can be exhaustive. As for the success rate, we disambiguate it in order to exclude unwanted sequences.

4 Extraction of Nominal Expressions

4.1 Lexical Modeling of Data

In order not to deal with all nominal expressions on a case-by-case basis, we were required to do data modeling. Modeling is a mold with all the properties that can cover all the expressions we want. Our modeling is based on observations of *Cuisitext* and dictionaries (*TLFI* and *Le petit Robert*). Since our study concerns the modeling of collocation, we also used lexical function [22], which is a "conceptual tool of language descriptions by modeling and encoding paradigmatic (semantic) and syntagmatic links" [22]. If we embody the lexical function by the mathematical formula: $f(x) = y$, x represents the argument (keyword) and y is its value. For example, the Lexical Function [*Prepar*] is a verb with the meaning "prepare for" which has the keyword L as a complement to a central object, and the Lexical Function [$Fact_0$] is a realization verb. Taking the case of "*ail*" as a keyword, this combination of the two lexical functions can be expressed as follows:

$$PreparFact0(ail) = hacher[\text{ART} \sim]$$

And the Lexical Function [A_2] is an adjectival modifier, so the adjective of the verb **P***reparFact0* (*ail*) can be manifested as follows:

$$A_2 PreparFact_0(ail) = hach\acute{e}$$

This FL helps to encode the semantic and syntactic links. On the semantic level, the lexie HACHÉ is the adjective of verb "*hacher*" which expresses the way to prepare, and on the syntagmatic level, this lexis is always the last ingredient that is going to be qualified. Thanks to the observation of our corpus, the expression "*ail éminée*" can be expressed in the form of two main structures: a nominal syntagm (ail haché): [Name + Adjective]; An attributive structure (ail est haché): [Name + Verb + Adjective].

Modeling of the Nominal Structure. For the first main structure [Name + Adjective], if there is no insertion between the name and the adjective of this structure, this expression can be constructed in *NooJ* as follows:

$$G = <N> <A>$$

If there is one or more insertions between the two elements, this expression can be constructed as follows:

$$G = <N> <WF> * <A>$$

The set of the two grammars can be described by the textual grammar:

$$G = <N> (<WF> * | <E>) <A>$$

Modeling of the Attributive Structure. As regards the second main structure [Name + Verb + Adjective], it can be transformed into other structures (L) and their corresponding grammars (G) of the structures are greatly varied and diversified to describe, as shown in Table 2.

Table 2. Data modeling

L=	G=
{"ail est écrasé"}	<N><être><A>
{"ail n'est pas écrasé"}	<N><ne><être><pas><A>
{"ail n'est pas assez écrasé"}	<N><ne><être><assez><A>
{"ail a été bien écrasé"}	<N><être><bien><A>
{"ail n'a pas été écrasé"}	<N><ne><avoir><pas><être><A>
{"ce sont ces ails qui doivent être écrasé"}	<ce><être><ce><ail><qui><devoir><être><A>
{"ces ails, ils doivent être bien écrasé"}	<ces><N>, <PRO><devoir><bien><être><bien><A>
{"ils doivent être écrasé, ces ails"}	<PRO><devoir><être><A>,<ces><N>
{"....................."}	....

It seems impossible to create so many modelizations for all the structures, but we have observed that the structures above share the same lexical materials: nominal group (GN), state verb (VE) and attribute (A). We can use transformational grammar that focuses on similar sentence relationships. From a sentence, the *NooJ* generator can produce all its transformational sentences and it can definitely recognize these transformed sentences. In a transformational analysis, we can use the global variables (prefixed by the "@" character) which have only one value, regardless of their place in the grammatical structure and this value is transmitted to the *NooJ* generator. We classify the GN in the subject (@S), the VE in the verb (@V) and the adjectives in the attribute (@A), from which the modelizations of the attributive structure are in Table 3:

Table 3. Modeling of the second main structure

G=	Phrases
@S(<WF> *|<E>)@V (<WF> *|<E>)@A	{"ail est écrasé"}; {"ail n'est pas écrasé"}; {"ail n'est pas assez écrasé"}; {"ail a été bien écrasé"}; {"ce sont ces ails qui doivent être écrasé"}
@S,<PRO>(<WF> *|<E>)@V (<WF> *|<E>)@A	{"ces ails, ils doivent être bien écrasé"}
<PRO> (<WF> *|< E>)@V (<WF> *|<E>)@A,@S	{"ils doivent être écrasé, ces ails"}

The three models can describe all transformed sentences which contain both GN as subject, VE as verb and adjectives as attribute in a sentence. The modelizations of the two main structures can extract not only phraseological and semi-phraseological expressions, but also free expressions. However, to reject free expressions, we need a disambiguation.

4.2   Disambiguation

NooJ offers us the operator (EXCLUDE) to reject unwanted sequences by imposing relevant constraints. For this reason, we have filtered all recognized expressions (phraseological, semi-phraseological and free expressions) by eliminating free expressions. Our constraints are based on both the names and adjectives of the recognized expressions. As regards the rejection of the adjective, we have gathered all free adjectives which are general adjectives existing in all domains under the label AL, such as <moyen>[1], <seul>, <entier>, <froid>, <inférieur>, <extérieur>, <dur>, <autre>, <fort>, <faible>, <différent>, <juste>, <extra>, <égal>, <nécessaire>, <normal>, <prêt>, <restant>, <suivant>, <régulier>, <allant>, <haut>, <petit>, <grand>, <joli>, <minimum>, <maximum>, <présent>; Similarly, regarding to the elimination of names, we have grouped

---

[1] In *NooJ*, the single quotation marks (<>) allow us to find all the occurrences of this term and its variants.

together all the free names which are the general names present in all domains under the NL label, for example, <la>, <être>, <avoir>, <bien>, <si>, <été>, <petit>, <y>, <g>, <tout>, <présent>, <courant>, <aide>, <dessus>,<ce>, <fois>, <type>, <préférence>, <fait>, <reste>, <les>, <le>, <intérieur>,<extérieur>, <morceau>. The proposed NL and AL are based on our observations on *Cuisitext*. Being limited to the inventory of names and general adjectives, our disambiguation remains restricted. Once the lexical modeling of the data and the disambiguation are finished, we begin their implementations in *NooJ*.

5   Implementation

5.1   Implementation of the Nominal Structure

For the implementation in *NooJ* (shown in Fig. 2), we use the variable N by imposing the constraints <$ THIS $ Nb = $ N $ Nb> and <$ THIS $ Genre = $ N $ Genre> to verify that the names correspond the adjectives in gender and number. The disambiguations (AL and NL rejected) are done both with the adjectives and the names.

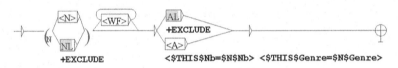

Fig. 2. Implementation of the nominal structure

5.2   Implementation of the Attributive Structure

The implementation of the attributive structure is shown in Fig. 3 as follows:

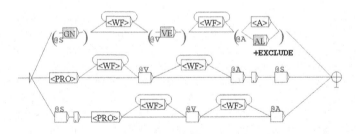

Fig. 3. Implementation of the attributive structure

We recall that the algebraic grammar allows a rule to be defined from itself. In Fig. 4 below, the nested graph of the nominal group (GN) refers to itself and this reference is recursive, because the GN is part of its own graph.

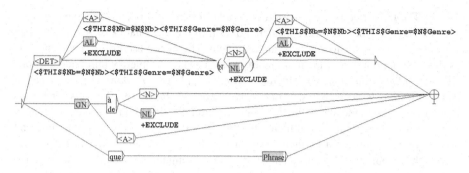

Fig. 4. Nested graph of the GN (Nominal Group)

In this nested GN graph, AL and NL rejected are always imposed for disambiguation and the variable N is also defined in gender and number.

Each main structure has its own graph: Fig. 2 for the nominal structure and Fig. 3 for the attributive structure. To solve this problem, we can merge the two graphs into a single graph in Fig. 5 which can make it possible to carry out the extraction in a single query.

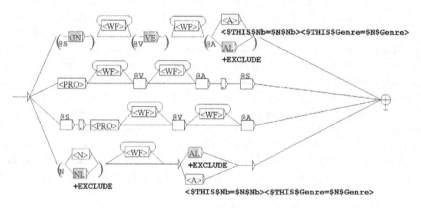

Fig. 5. Total graph

After completing the implementation of the data in *NooJ*, we launch our request.

6 Qualitative and Quantitative Analysis of the Results

Our corpus is too large to perform a syntactic and semantic analysis in *NooJ*, only part of the corpus (50 thousand words) is taken into account for this extraction. Thanks to the total graph, we have obtained a total perception of the result of the extraction. As a result of our study, 353 extracted sequences are ranked in a decreasing order of occurrences of words and 338 sequences are retained. The success rate of extraction is

95.7%. The most frequent expression is *"crème fraîch"* (38 occurrences) and it turns out that *"crème fraîch"* is not considered a term in the dictionaries of *NooJ*, although *"pomme de terre"* is listed there as an ingredient. The noises according to which some expressions are not part of the nominal expressions were completely unfounded: *h sous un linge humide dans un endroit tiède* (Rank 46), *accompagnant le magret coupé* (Rank 50), *lait de coco* (Rank 34) et *thon au naturel* (Rank 43). The answer to this problem is found in the annotations of our extraction: the first and last words of the four expressions are their names and their adjectives. All the elements between the first and last words are considered as their insertions (WF). In such a case, in fact, we have solved this problem by taking advantage of the disambiguation of the names or the adjectives. For example, the additions of *h* and *accompanying* in the rejected NL will eliminate the first two extracted expressions. This means that the disambiguation improves, to a certain extent, the result of the extraction.

7 Conclusion

The salient features of our corpus (*Cuisitext*), written corpus and oral corpus both included, are expressions of the phraseology. Our simple linguistic analysis shows us that the crucial problem of phrase extraction is that of semi-phraseological expression. A search in the TAL domain is done in order to obtain suitable software of the phraseological extraction. *NooJ*, as a corpus processing system, can perfectly fulfill our two requirements: disambiguation and grammatical transformation. Modeling and disambiguation are at the heart of extraction. Our encodings based on observation and lexical function allow us to easily recover some irregular expressions, such as *le caramel, qui ne doit pas devenir noir* (Rank 24), *crème fraîche liquide* (Rank 26), *la pâte soit plus ferme* (Rank 32), *fromage de chèvre frais* (Rank 49), etc. Due to the qualitative and quantitative analysis of the extraction result, we are aware that the adjustment of the disambiguation of nouns and adjectives can improve the success rate of the phrase extraction result. After having succeeded in extracting the nominal expressions from the phraseology, a selection of expressions to be taught will be made in the continuation of our thesis: the teaching of the nominal expressions of the phraseology with the foreign learners.

References

1. Anthony, L.: AntConc: design and development of a freeware corpus analysis toolkit for the technical writing classroom. In: Communication présentée à la Conference IPC 2005, pp. 729–737 (2005)
2. Cavalla, C.: La phraséologie en classe de FLE. Les Langues Modernes 1/2009 (2009). http://www.aplv-languesmodernes.org/spip.php?article2292
3. Cavalla, C., Loiseau, M.: Scientext comme corpus pour l'enseignement. In: Tutin, A., Grossmann, F. (Eds.) L'écrit scientifique: du lexique au discours. Autour de Scientext, Rennes: PUR, pp. 163–182 (2013)
4. Church, K., Gale, W., Hanks, P., Hindle, D.: Using statistics in lexical analysis, pp. 115–164 (1991)

5. Church, K.W., Hanks, P.: Word association norms, mutual information, and lexicography. In: Proceedings of the 27th Annual Meeting of the Association for Computational Linguistics, 26–29 June 1989, Vancouver, Canada, pp. 76–83 (1989)
6. Chomsky, N.: Syntactic Structures. Mouton, The Hague (1957). Livre traduit en 1969: Structures syntaxiques. Le Seuil, Paris
7. Choueka, Y., Klein, S.T., Neuwitz, E.: Automatic retrieval of frequent idiomatic and collocational expressions in a large corpus. J. Assoc. Lit. Linguist. Comput. **4**, 34–38 (1983)
8. Cowie, A.P.: The place of illustrative material and collocations in the design of a learner's dictionary. In: Strevens, P. (ed.) In Honour of A.S. Hornby. Oxford University Press, Oxford, pp. 127–139 (1978)
9. Goldman, J.P., Nerima, L., Wehrli, E.: Collocation extraction using a syntactic parser. In: Proceedings of the ACL 2001 Workshop on Collocation, Toulouse, pp. 61–66 (2001)
10. González-Rey, I.: La phraséologie du français. Presses Universitaires du Mirail, Toulouse (2002)
11. González-Rey, I.: La didactique du français idiomatique. E.M.E., Fernelmont (2008)
12. Grefenstette, G., Teufel, S.: Corpus-based method for automatic identification of support verbs for nominalizations. In: Proceedings of the Seventh Conference of the European Chapter of the Association for Computational Linguistics, 27–31 March 1995, Dublin, Ireland, pp. 98–103 (1995)
13. Gross, M.: Une classification des phrases «figées» du français. Revue québécoise de linguistique **11**(2), 151–185 (1982)
14. Grossmann, F., Tutin, A.: Les collocations: analyse et traitement. De Werelt, Amsterdam (2003)
15. Kilgarriff, A., Tugwell, D.: Word sketch: extraction, combination and display of significant collocations for lexicography. In: Proceedings of the Workshop on Collocations: Computational Extraction, Analysis and Exploitation, ACL-EACL 2001, Toulouse, pp. 32–38 (2001)
16. Lamiroy, B., Klein, J.R.: Le problème central du figement est le semi-figement. Linx **53**, 135–154 (2005)
17. Lewis, M.: Teaching Collocation, Further Developments in the Lexical Approach. Language Teaching Publications LTP, Hove (2000)
18. Lin, D.: Extracting collocations from text corpora. In: First Workshop on Computational Terminology, Montréal, pp. 57–63 (1998)
19. Luka, N., Seretan, V., Wehrli, E.: Le problème de collocation en TAL. In: Nouveaux cahiers de linguistiques Française, pp. 95–115 (2006)
20. Mejri, S.: Figement, néologie et renouvellement du lexique. Linx. Revue des linguistes de l'université Paris X Nanterre **52**, 163–174 (2005). https://doi.org/10.4000/linx.231
21. Mel'čuk, I.: La Phraséologie et son rôle dans l'enseignement-apprentissage d'une langue étrangère. Études de linguistique appliquée **92**, 82–113 (1993)
22. Polguère, A.: Towards a theoretically-motivated general public dictionary of semantic derivations and collocations for French. In: Proceedings of EURALEX 2000, Stuttgart, pp. 517–527 (2000)
23. Polguère, A.: Lexicologie et sémantique lexicale: notions fondamentales. Troisième édition (première édition en 2003), les presses de l'Université de Montréal, Montréal (2016)
24. Sag, I., Baldwin, T., Bond, F., Copestake, A., Flickinger, D.: Multiword expressions: a pain in the neck for NLP. In: Proceedings of the Third International Conference on Intelligent Text Processing and Computational Linguistics (CICLING 2002), Mexico City, pp. 1–15 (2002)
25. Salem, A., équipe SYLED: Statistique textuelle. Dunod, Paris (2001)

26. Silberztein, M., Tutin, A.: NooJ, un outil TAL pour l'enseignement des langues: application pour l'étude de la morphologie lexicale en FLEM. Alsic, **8**(2), 123–134 (2005)
27. Smadja, F.: Retrieving collocations form text: Xtract. Comput. Linguist. **19**(1), 143–177 (1993)
28. Tutin, A.: Pour une modélisation dynamique des collocations dans les textes. Actes d'Euralex, Lorient (2004)
29. Yang, T.: Cuisitext: un corpus écrit et oral pour l'enseignement, colloque LOSP (Langues sur objectifs spécifiques: perspective croisées entre linguistique et didactique), Grenoble, 24–25 November 2016 (2016). http://losp2016.u-grenoble3.fr

Sentiment Analysis Algorithms for the Belarusian NooJ Module in Touristic Sphere

Yuras Hetsevich[1], Alena Kryvaltsevich[1(✉)], Nastassia Kazloŭskaja[1],
Anastasija Drahun[1], Jaŭhienija Zianoŭka[1],
and Aliaksandr Ščarbakoŭ[2]

[1] The United Institute of Informatics Problems,
National Academy of Sciences of Belarus, Minsk, Belarus
yuras.hetsevich@gmail.com, alena.ssrlab@gmail.com,
krasnova.an.23@gmail.com, ndrahun@gmail.com,
evgeniakacan@gmail.com
[2] The Belarusian State University of Informatics and Radioelectronics,
Minsk, Belarus
alexandrscherbakov231194@gmail.com

Abstract. Sentiment analysis is the area of computational linguistics that investigates a statistical probability of an emotional component in a text or speech. Sentiment analysis is often used in such spheres as social media and tourism. The main task of the analysis is to find the keywords of opinion in the text and to define their properties depending on the task, for example, who owns this opinion, the topic of the opinion and the tone (positive, negative or neutral). As sentiment analysis algorithms for the Belarusian language is undeveloped sphere, the authors have decided to model this mechanism in NooJ as a linguistic processor. The authors have chosen the touristic sphere as it is a highly developing branch of the Belarusian state economy. We are developing sentiment analysis algorithms in the borders of touristic domains texts, based on opinion mining for Belarusian cities. Results of this research are to enlarge resources for the Belarusian NooJ module, enable to study new levels for further research in other domains and can be used for solving different linguistics and sociological tasks.

Keywords: Sentiment analysis · Touristic sphere · Opinion
The Belarusian NooJ module · The Russian language · Mirski castle
UNESCO heritage

1 Introduction

Sentiment analysis is a process of the emotional polarity extraction. It is not about only positive or negative emotions, but mostly, sentiment analysis aims to determine the opinion of a speaker, writer, or another subject with respect to some topic; sentiment analysis is a complex area, which tries to apply statistical, computational and objective methods to subjective manifestations.

© Springer International Publishing AG 2018
S. Mbarki et al. (Eds.): NooJ 2017, CCIS 811, pp. 179–189, 2018.
https://doi.org/10.1007/978-3-319-73420-0_15

Sentiment analysis is important for numerous scientific and industrial purposes. The main aim of sentiment analysis algorithms is to detect an emotional component in speech or text. The basic task of sentiment analysis is to define the polarity of a given text. The study in this field started approximately in the 1990s [1, 2]. However, most of the related works on sentiment analysis [1, 3] have focused on the identification of sentiment expressions and their polarities [4].

Sentiment analysis techniques may be used as a component of greater systems, for example, in marketing and user's behavior predicting systems. Customers are usually happy to exclude from the recommendations the items that have received very negative feedback.

The design of automatic tools that are capable to mine sentiments over the Web in real-time and to create condensed versions of found information is one of the most active research and development areas [5].

In this paper, we cover the topic of semantic analysis in a touristic sphere. Touristic sphere is an important part of life not only as a state income, but also as a mean for developing someone's personality. According to the State Program for the Development of Tourism in the Republic of Belarus for 2016–2020 [6], our country will be more open to a touristic flow to increase interest of foreign tourists.

Additionally, Belarusian team have developed Belarusian NooJ module since 2011. We have started to create the first Belarusian, Russian *.nod dictionaries, publishing of Belarusian NooJ module with texts [10], projects [11], grammars [12, 13], dictionaries [14]. In addition, we have started new Java NooJ core implementations.

So in this paper, we have included the results of the touristic domain investigations in terms of UNESCO heritage object—Mirski castle, as well as opening the topic of Java NooJ creation as a core for the implementation of developed Belarusian NooJ modules online. In this paper, you may find the following terms:

– Experiment—an application of NooJ syntax grammar to the corpus of manually collected texts.
– Class—an extensible program-code-template for creating objects, providing initial values for state (member variables) and the implementations of behavior (member functions or methods) [7].
– Feature—an expression that contains a particular sentiment component (positive or negative).

2 An Overview of the Touristic Domain in Belarusian Information Space

The development of Belarusian information space has started since 1994. Now there are more than 70000 sites in the domain name's label ".by". Some of these resources are http://otzyvy.by/ (reviews on goods and services) and http://orabote.by/ (reviews on job and jobseekers). You may find some reviews on the particular site of the museum or a hotel, but, unfortunately, not every museum in Belarus has its site.

In our investigation, we have concentrated on purely positive and negative reviews on Mirski castle (in Belarusian, Мірскі замак). We chose this particular cultural object because it is one of the UNESCO heritage objects and it is rather popular among visitors from abroad and inside our country (Fig. 1).

Fig. 1. The photo of Mirski castle

As the corpus of reviews in the Belarusian language will not be representational enough, we decided to collect a corpus in the Russian language. In our selection, there are reviews that are written not only by native Belarusians but also by foreign visitors. Important criteria—the foreign visitor that left a review should be a native speaker of the Russian language.

All reviews could be found in the open Internet resources (mostly from https://www.tripadvisor.com/). All reviews were collected manually.

The main task of this research was to work out a preliminary algorithm of the sentiment analysis for the Russian language. It is worth saying that for each of the experiment we attempted to find a better solution of sentiment analysis problem. So we modified algorithm three times and this algorithm was externated in NooJ syntax grammars and improved corpora. We had a new evaluation for each of the experiment.

In future, we are planning to annotate adjectives and adverbs according to their sentiment of all words extracted from Belarusian corpus with adding such annotated words to the present dictionary based on the Belarusian NooJ module.

3   Corpora Collecting

Our team collected all the reviews manually using the popular foreign and local resources, such as "tripadvisor.com", "mircki-zamok.relax.by", "forum.onliner.by", etc.

The example of positive review is "*Замок **очень красивый** и внутри,и снаружи. **Впечатляющие** виды днем и вечером с подсветкой. Рядом озеро и небольшой парк с усыпальницей*". (In English, it could be translated as "*the castle is very beautiful inside and outside. Impressive views during a day and in the evening with tap lights. Nearby there is a lake with small park and table-tomb*").

The example of negative review is "*При въезде в город сам замок **не виден**, указателей **никаких нет**. Немножко **заплутали**, но благо город небольшой,* *поэтому быстро нашли. Конечно замок вживую* **не производит величест** **венного впечатления.** *Но при покупке билета нам* **не было предложено** *аудиогида. (Обязательно просите, потому что без него или без экскурсовода там* **нечего делать***). Экспозиция* **очень слабенькая***".* (In English, it could be translated as "*At the entrance to the city the castle itself is not visible, there are no signs. We strayed a little bit, but the city is rather small, so we quickly found our way. Of course, the castle does not produce a majestic impression alive. However, while buying a ticket, taking an audio guide was not offered. (Be sure to ask, because without it or without a guide there is nothing to do). The exposition is very weak*").

In the examples above, we marked with bold type the lexemes and phrases that definitely point at the review positivity or negativity. These features contain different parts of speech and modifiers, but we left the analysis of these features for our further investigations.

4 Experiments

We have chosen a rule-based method. As a particular rule, we used rigidly fixed expressions that include definitely clear positive or negative connotation (sentiment). As for the materials for the experiments, we used reviews on Mirski castle that were collected manually from different open sources. The reviews were written by native speakers in Russian.

Concerning tools, we used NooJ [8, 9] as an instrument of investigation, and NooJ syntax grammar as a tool for experiments. In the graphs of syntax grammar, we placed particular emotional features (i.e. phrases) extracted from reviews.

Conducting the experiments we tried to achieve the maximum rate of F-measure and tried to check, whether it is possible to apply a rule-based method that has a strictly limited list of positivity-negativity features as rules successfully (with a rate of F-measure not lower that 70%).

4.1 First Experiment

Review corpus for the first experiment contains 503 positive reviews and 168 negative reviews (Fig. 2). They were collected from the following resource: https://www.trip advisor.ru/Attraction_Review-g1075850-d319498-Reviews-Mir_Castle-Mir_Grodno_ Region.html

In addition, we extracted particular features that definitely describe emotion: 404 positive, 348 negative.

The corpus for the first experiment contained 503 positive reviews, 168 negative reviews, as well as 404 positive and 348 negative features in syntax grammar subgraphs. We mastered our naive positive-negative algorithm in NooJ with the help of this corpus (Fig. 3).

We used subgraphs of NooJ syntax scheme as corpora for featured rules (i.e. emotional features) (Fig. 4). Under featured rules or features, we understand manually collected and analyzed lexemes and phrases that may definitely testify a review positivity or negativity.

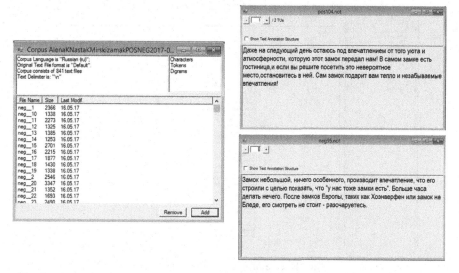

Fig. 2. Corpus for the first experiment with examples of reviews (positive and negative)

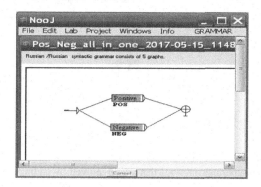

Fig. 3. Naive positive-negative algorithm

The featured rules were extracted from the texts that were used for this first experiment. Then, we improved our grammar (Fig. 5) with adding the negative particle "*не*" (in English it can be equivalent to "*not*") as an additional unit. This step widened our algorithm with such parameters: positive features (they get tag "POS"), non-negative features (they get tag "POS"), non-positive features (they get tag "NEG"), and negative features (they get tag "NEG"). For example, in positive features' corpus there is a word "*хорошо*" (in English it is translated as "*good*"); it will get a POS-tag. But the phrase ["*не*" + "*хорошо*"] (in English it could be translated as "*not good*") should be tagged as NEG, because particle "*не*" is a negative particle. We decided to use the possibilities of NooJ syntax grammar to solve the problem of such phrases tagging.

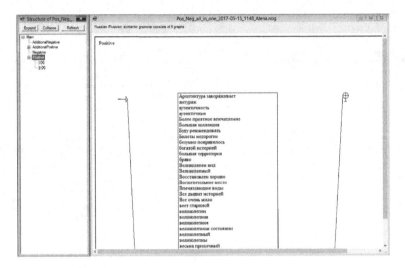

Fig. 4. One of the "positive" subgraph

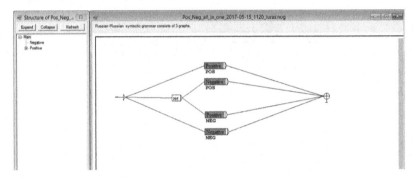

Fig. 5. Second syntax grammar in NooJ that was improved by adding "*не*" negative particle

Second syntax grammar (Fig. 5) analyzed 503 positive and 168 negative reviews. We got 91% of precision, 16% of recall, 27% of F-measure (first evaluation). After the experiment, we decided to raise the level of recall and F-measure with the enlargement of featured rules list, as well as with review corpus widening, because we have proved the main algorithm concept (Fig. 6).

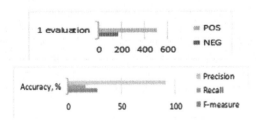

Fig. 6. First experiment results' charts

4.2   Second Experiment

To make the second experiment on the widened corpora of texts, we added to the first experiment's review corpus, 131 positive reviews and 39 negative reviews. We got the second experiment's corpus that consists of 634 positive and 207 negative reviews for the second experiment.

The features' corpus for the second experiment includes the features' corpus for the first experiment and 524 positive, 98 negative features, so that we have got the features' corpus with 928 positive and 446 negative features.

The first experiment gave us 91% of precision, 16% of recall and 27% of F-measure. For the second corpus, we collected 131 positive and 39 negative review, and extracted 524 positive and 98 negative features. After that, we concatenated the first and second review corpora and the features' corpora; the result of it was the concordance of tagged features. As an example of the concordance with POS and NEG tag you may see the concordance for the first experiment (Fig. 2).

In total, we have got 634 positive and 207 negative reviews and 928 positive and 446 negative features for the second experiment.

Additionally, for the second experiment we deepened the structure of the syntactic grammar to get higher scores (Fig. 7) with additional 131 positive and 39 negative features.

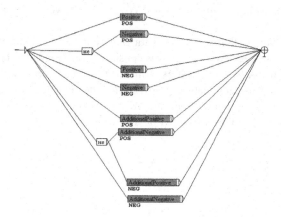

Fig. 7. Experimental grammar based on the corpus for the second experiment

For the second experiment, we used 634 positive and 207 negative reviews and 928 positive and 446 negative features in syntactic grammar graphs (Fig. 8).

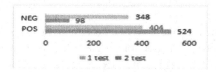

Fig. 8. Comparison of featured rules' amount for the first and second experiment

As for the result of our experiment, we obtained 92% of precision, 89% of recall, 90% of F-measure (Fig. 9).

Fig. 9. Second experiment results' charts

After gaining the above-mentioned results, we decided to hold a final experiment with the use of unanalyzed text corpus. Analyzed text corpus is a corpus from which we did not extract emotional features.

4.3   Final Experiment

The final experiment corpus contains 83 positive reviews and 24 negative reviews. They were collected from open resources.

There were no features extracted from the above mentioned reviews to prevent inaccuracy in the calculations of precision, recall and F-measure. In the syntactic grammar that we applied to this final test corpus, there were 928 positive and 446 negative features.

The features' corpus that was created for the second experiment, which contained 928 positive and 446 negative features was considered as a gold-standard collection, as a right variant. Finally, we did a test experiment on a collection of reviews, which we did not analyze and use for features' extraction. The final test corpus contained 83 positive and 24 negative reviews. After the final experiment, we got 94% of precision, 70% of recall and 80% of F-measure (Fig. 10).

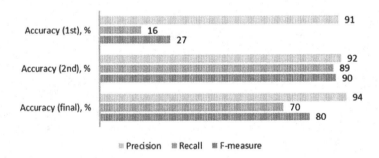

Fig. 10. Charts of accuracy improving during the first, second, and final experiments

Therefore, we created a rule-based algorithm with the help of NooJ syntactic grammar using its graphs as a tool for tagging the emotional features in the text. Finally, we obtained 94% of precision, 70% of recall and 80% of F-measure on the corpus of unanalyzed reviews.

5 Additional Investigations

Our team developed new dictionary module for NooJ as well as new RegExp module and tested it under the test set that includes short text, and six tasks for NooJ Linguistic Analysis and Locate pattern functionality of NooJ formalism.

We used C# NooJ version as a working program and created an open source of Java NooJ from meta-share for modification. Our investigation in this direction was based on NooJ Manual [8] as a handbook of NooJ formalism. We used sample files of *.dic, *.nof, properties from the Belarusian NooJ module and Model-View-Controller approach.

We developed next classes (Table 1):

Table 1. Java NooJ classes description

Class	Description of class
AnalisedWord	implements full linguistic information. This class contains the following fields: word form, particular forms of words in the text, and other linguistic information
LexicalInformation	implements a data model and stores the basic linguistic information word as described in the dictionary. It includes fields such as word form, part of speech, and other linguistic information
Paradigm	implements a data model for storing in the computer memory all inflected grammar words. It includes fields such as word form, the number of changes, the symbol at the end of the word, and the variable part of the word
Property	implements a data model and stores the additional linguistic information. It includes information about the part of speech and other linguistic information
SercherOutputPrinter	implements a data model and stores the context in which the word is used in the text. It includes fields such as the context before the word, the word, and the context after the word
Gramm	stores in the computer memory all inflected grammar words. It includes fields such as the word form, all forms of words with corresponding linguistic information
DefWorker	provides methods that implement the file processing function of the type «Dictionary properties definition»
DicWorker	provides methods that implement the file processing function of the type «Dictionary»
NofWorker	provides methods that implement the file processing function of the type «Inflectional/Derivational description»
TextWorker	provides methods that implement the file processing function of the text type files
Analysis	provides an implementation of secondary processing of the text prior to analysis, the construction of grammars, text analysis, and the preparation of the analysis results to the output file, and the output of the analysis of the text in the file

In future, we will be able to implement constructed algorithms of the sentiment analysis with the Belarusian NooJ module for tracking the opinion of our users about our national sightseen in particular systems. We hope to develop the web and mobile applications that will use developed algorithm, which is based on NooJ syntactic grammars.

6 Conclusion

Summarizing our research, the following results should be underlined. Firstly, there is no open-source technical decision for the Belarusian language sentiment analysis for the touristic sphere. Mostly it could be explained by the difficulty to find text content in the Belarusian language, especially on the touristic topic. That is why Belarusian NooJ team has chosen texts of reviews on Belarusian sightseen but in the Russian language. Secondly, NooJ software can be successfully implemented to help in the basic sentiment analysis investigations. Our team developed a rule-based algorithm that showed 94% of precision, 70% of recall and 80% of F-measure during the final experiment.

In the sphere of the sentiment analysis, Belarusian NooJ team plans to improve grammars, in particular, to add <WF> and <P> modifications into grammars for more accurate work of grammars in a general case. In addition, our team is going to improve the constructed syntactic grammars algorithm by adding Adverbs and Adjectives to the subgraphs, and a tag for the neutral state. Additionally, Belarusian NooJ team plans to do modifications in Excel calculations to handle the exceptions such as multiple POS (NEG) annotations for the same review.

We are going to build an online Java Belarusian NooJ prototype for the sentiment analysis automation. In addition, Belarusian NooJ team plans to develop new score of Java NooJ, in particular, complete RegExp module, finalize the dictionary module, and extend functionality.

References

1. Hatzivassiloglou, V., McKeown, K.R.: Predicting the semantic orientation of adjectives. In: Proceedings of the 35th Annual Meeting of the ACL and the 8th Conference of the European Chapter of the ACL. Association for Computational Linguistics, Madrid, Spain, pp. 174–181 (1997)
2. Baker, C.F., Fillmore, C.J., Lowe, J.B.: The Berkeley FrameNet project. In: Proceedings of the Joint Conference on Computational Linguistics and the 36th Annual Meeting of the ACL (COLING-ACL98). Association for Computational Linguistics, Montreal, Canada (1998)
3. Hatzivassiloglou, V., Wiebe, J.M.: Effects of adjective orientation and gradability on sentence subjectivity. In: Proceedings of 18th International Conference on Computational Linguistics (COLING), pp. 299–305 (2000)
4. Nasukawa, T., Yi, J.: Sentiment analysis: capturing favorability using natural language processing, pp. 70–77 (2003)
5. Cambria, E., Dass, D., Bandyopadhyay, S., Feraco, A.: A Practical Guide to Sentiment Analysis, pp. 1–2. Springer, Heidelberg (2017). https://doi.org/10.1007/978-3-319-55394-8

6. State program on tourism developing in Republic of Belarus (2017). http://www.mst.by/ru/programma-razvitiya-turizma-ru/. Accessed 17 Jan 2017. (Electronic resource)
7. Bruce, K.B.: 2.1 Foundations of Object-oriented Languages: Types and Semantics, MIT Press, Cambridge, p. 18 (2002)
8. Silberztein, M.: NooJ manual. www.nooj4nlp.net (2003)
9. Silberztein, M.: Formalizing Natural Languages: The NooJ Approach. Wiley, London (2016)
10. Reentovich, I., Hetsevich, Y., Voronovich, V., Kachan, E., Kozlovskaya, H., Tretyak, A., Koshchanka, U.: The first one-million corpus for the Belarusian NooJ module. In: Okrut, T., Hetsevich, Y., Silberztein, M., Stanislavenka, H. (eds.) NooJ 2015. CCIS, vol. 607, pp. 3–15. Springer, Cham (2016). https://doi.org/10.1007/978-3-319-42471-2_1
11. Borodina, J., Hetsevich, Y.: Using NooJ to process satellite data. In: Okrut, T., Hetsevich, Y., Silberztein, M., Stanislavenka, H. (eds.) NooJ 2015. CCIS, vol. 607, pp. 182–190. Springer, Cham (2016). https://doi.org/10.1007/978-3-319-42471-2_16
12. Hetsevich, Y., Okrut, T., Lobanov, B.: Grammars for sentence into phrase segmentation: punctuation level. In: Okrut, T., Hetsevich, Y., Silberztein, M., Stanislavenka, H. (eds.) NooJ 2015. CCIS, vol. 607, pp. 74–82. Springer, Cham (2016). https://doi.org/10.1007/978-3-319-42471-2_7
13. Lysy, S., Stanislavenka, H., Hetsevich, Y.: Addition of IPA transcription to the Belarusian NooJ module. In: Barone, L., Monteleone, M., Silberztein, M. (eds.) NooJ 2016. CCIS, vol. 667, pp. 14–22. Springer, Cham (2016). https://doi.org/10.1007/978-3-319-55002-2_2
14. Hetsevich, Y., Varanovich, V., Kachan, E., Reentovich, I., Lysy, S.: Semi-automatic part-of-speech annotating for Belarusian dictionaries enrichment in NooJ. In: Barone, L., Monteleone, M., Silberztein, M. (eds.) NooJ 2016. CCIS, vol. 667, pp. 101–111. Springer, Cham (2016). https://doi.org/10.1007/978-3-319-55002-2_9

Question-Response System Using the NooJ Linguistic Platform

Imen Ennasri, Sondes Dardour, Héla Fehri, and Kais Haddar[(⊠)]

MIRACL Laboratory, University of Sfax, Sfax, Tunisia
imen.ennasri@yahoo.fr, dardour.sondes@yahoo.com,
hela.fehri@yahoo.fr, kais.haddar@yahoo.fr

Abstract. Nowadays, the access to medical information is a crucial task given the large number of electronic documentation and its various sources, particularly on the internet. Search engines such as Google, Yahoo, etc. establish an effective solution to find documents corresponding to a user request but they provide imprecise and fast information corresponding to user needs. This research gap motivated us to develop a question-answer system allowing users to ask a question about the desired information using natural language without browsing through documents. The request is presented as few key words; our system responds by a precise and quick answer thanks to the various features provided by the development environment NooJ. The maturity and the efficiency of the search medical information tools are classified according to the level of complexity of the subject area and the target language. Despite of several researches, tools for Arabic language remain relatively lacking many features compared to other languages. The implementation of the proposed system is based on two processes: the first process consists of identifying the type and keywords of the question with the aim of limiting the number of responses. The second process consists of applying the appropriate transducer set to the corpora to extract responses.

Keywords: Medical information research · Question-answer system
NooJ · Arabic language · Transducer

1 Introduction

The medical field now has a massive amount of electronic documents enabling the search for any medical information. Indeed the search for precise and complex medical information in terms of time [1]. Arabic and English are rich of complex structures, especially Arabic that is an agglutinated, strongly inflected, and derivational language. We need to build a system that studies the Arabic and English medical corpora, which facilitates answers extraction and largely reduces the research time.

The aim of this paper is to propose a Question-Answer System providing information related to the medical domain (e.g. disease, drug). The question is introduced in Arabic and the response could be displayed in Arabic and in English. In order to develop such a system, we were confronted with several issues. For instance, the response corresponding to a given question could be ambiguous. For this reason, it is important to precisely define different keywords in order to avoid the confusion between questions and responses.

S. Mbarki et al. (Eds.): NooJ 2017, CCIS 811, pp. 190–199, 2018.
https://doi.org/10.1007/978-3-319-73420-0_16

The implementation of the proposed system is based on two processes: the first process consists of identifying the type and keywords of the question with the aim of limiting the number of responses. The second process consists of applying the appropriate transducer set to the corpora to extract responses. This step is based on bilingual dictionaries that contain the different Arabic keywords related to the medical domain and their translation into the English language. In fact, the developed system should search for the response in the Arabic and the English corpora. The request form is developed by java programming language. The content of this form is saved as text that is an input of the linguistic platform NooJ. The given results are displayed in a Java interface in Arabic and English.

Indeed, this type of system allows the user to ask a question in Arabic and receive a precise answer to his request instead of a set of documents deemed relevant, as is the case for search engines.

In this work we begin by presenting a state-of-the-art study on current Question/Answer systems, the approaches, characteristics and strengths of each model. Next, we show the overall architecture of the proposed question-response system. After we build an Arabic-English corpus, analyze issues, build the necessary dictionaries and recognize medical entities through the platform NooJ, the experimentation and evaluation take place.

2 Question/Answer Systems

The Question/Answer systems exploring new methods of search information exploiting queries formulated using natural language and not based only on keywords (like current search engines) [2]. The Question/Answer system then uses automatic language processing techniques to analyze the issue and search for an adequate response to the documents which it has access to.

In proposing a suite of documents according to their estimated interest, the conventional search engine method forces the user to make a post-sort of relevant documents by himself. Many of the proposed pages that do not meet the question, being sometimes inconsistent, spread over different pages, etc. In the case of Question/Answer system issues, we generally try to reconstruct a natural language answer and not to offer the user a (long) list of documents.

The Question/Answer systems have three main goals [3]:

- Understanding the questions in natural language (analyzing the question, what is the type, what is its domain. We speak of closed answer to a question on a specific domain (medicine, computing…), and open issues that may deal with anything and for which there may be an appeal to the general ontology systems and knowledge about the world.
- Finding the information (either in database structured (database specialized) or in heterogeneous texts (Web search)).
- Answering the question (either by an exact answer, or the proposal of passages that may contain the answer).

Question-response systems can be differentiated according to the research strategies used. In the following table, we present some characteristic approaches that have generated the best results in question-response tasks during the recent TREC, CLEF and EQueR evaluation campaigns. The following two tables show the characteristics of each system, as well as the methods used [4] (Table 1).

Table 1. Comparison of systems

SQR	Approaches	Characteristics
QALC	Terminology indices	- first SQR developed in English
		- relies on a set of automatic language processing modules that operate downstream of a search engine operating on a wide selection of documents
		- performs a first processing on the question, performed by a dedicated partial parser, which determines the type of the expected response, the category of the question, the named entities of the question and the focus of the question
QRISTAL	TAL	- extracts responses from a local document database or from the web
		- consists of several modules for automatic language processing (syntactic analysis, semantic disambiguation, a search for the referents of anaphores…)
		- multilanguage (FR, Ang, Por, It, Pol)
		- characterized by an indexing motor
PIQUANT	Statistical	- uses several question-answer systems depending on the type of question which relies on different independent agents to search for an answer (among them an agent based on statistical tools and TAL)
		- a syntactic analysis to determine the type of question and desired response, keywords and a semantic form of question
		- Uses several knowledge sources such as WordNet, CYC
		- Better relevance thanks to the plurality and redundancy of the answers found
JAVELIN	Interaction with the user	- determines certain characteristics of the question: its type according to a predefined classification specific to the system, the type of the expected response, the keywords with their variants thanks to the WordNet semantic network
		- provides the user with a justification of the response by adding the description of the treatments performed by the system
PowerAnswer	Logical reasoning	- represent the question in logical formulas of the question, the answer as well as the data sources used to extract the answer

(continued)

Table 1. (*continued*)

SQR	Approaches	Characteristics
		-Using WordNet semantic data for analysis
		- uses a named entities recognition module
		-using an automatic response demostration program
		- return to the user not only the passages containing the answer but also the chain of reasoning linking the question and the answer
WEBCOOP	Inferences	-provide the user with additional information (explanations, justifications, etc.)
		- uses an ontology of the domain and knowledge bases combining the aspects of hosting and transport
InsigntSoft	Extraction patterns	- Analysis of questions by selecting of patterns, each pattern being applied to all candidate passages
		- matching is done on the basis of the pattern and elements of knowledge base (countries, currencies, etc.)

Each system has a different approach, uses different methods and also has specific characteristics. However, the main objective of these systems is to find a quick and effective response. So, the evaluation of the different systems depends on the accuracy of the obtained answer.

Table 2. Evaluation of systems

	Indexing	Multilingual	Syntactic analysis	Semantic analysis	Justification for response
QALC	–	–	+	–	–
QRISTAL	+	+	+	+	–
PIQUANT	–	–	+	+	–
JAVELIN	–	–	–	+	+
PowerAnswer	–	–	–	+	–
WEBCOOP	–	–	–	–	+
InsigntSoft	–	–	–	+	–

Campaigns for evaluating question-response systems, such as the TREC campaign, demonstrated that systems using semantic resources achieved the highest scores [5]. Moreover, and despite of the fact that the selection of keywords is the most important process for a question-response system but it remains difficult to achieve.

Although the performance of question-answer systems depends on their ability to find answers in the documents, it also depends heavily on the results returned by the search engines. As Table 2 shows, the QRISTAL system seems to be the best system but it has a great weak point, which is the justification for response.

For these reasons, we have chosen to work with the NooJ linguistic development environment. The latter is a corpus processing system that allows constructing, testing and managing formal descriptions in large coverage of natural languages, in the form of electronic dictionaries and grammars. It also allows the construction, editing and sophisticated management of concordances.

3 Proposed Method

In this section, we describe the overall architecture of the proposed question-response system. The latter contains four steps: the building of Arabic-English corpora, the question analysis, the construction of necessary dictionaries, and the recognition of medical entities. Figure 1 shows the different steps of the proposed method.

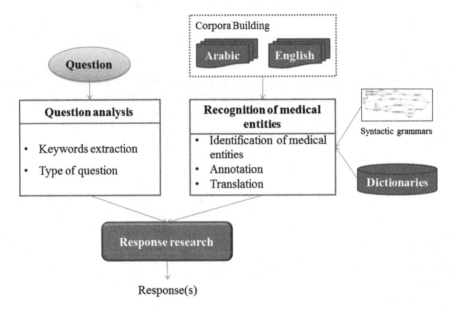

Fig. 1. The overall architecture of the question-response system

In the following section, we provide detailed examples about these steps.

3.1 Questions Analysis

Concerning the process of questions and their corresponding answers, we define four question types as follows:

Definition of Diseases. This type of question allows the respondent (system) to define the disease. For example: ما هو مرض القلب (What is heart disease?).

Symptoms of Disease. This type of question requires the respondent (system) to cite the symptoms that identify such disease. For example: ما هي اعراض التهاب المعدة (What are the symptoms of gastritis?).

Causes Disease. Causes disease question allows the respondent (system) to select infections, injury or the unhealthy lifestyles that cause ill health. For example: مرض الكبد ما اسباب (What are the causes of liver disease?).

Types of Disease. Types of disease question requires the respondent (system) to cite the different types of disease. For example: ما هي انواع امراض المعدة (What are the types of stomach diseases?).

The keyword of the question is used to extract pertinent answers. Indeed, to elicit keywords, we first collect « named entities ». In medical domain, the named entities are mainly technical names: diseases, symptoms, treatments, etc. The extracted Arabic keywords must be translated to extract English responses [6].

3.2   Corpora Building

We generate a corpus for each question type (Arabic and English). Each corpus regroups a number of texts. These corpora allow us to identify rules and transform them later into transducers.

3.3   Construction of Necessary Dictionaries

For our method, we construct three dictionaries. One contains the diseases' names. The second dictionary contains trigger words. This trigger generally defines the question type such as the word "اعراض" that define the question type "symptoms of disease". The last dictionary contains organs. Table 3 shows the number of entries of each dictionary.

Table 3. Dictionaries

Dictionaries	Entries
Diseases	20
Triggers	50
Organs	100

3.4   Recognition of Medical Entities

A set of patterns are modeled and translated into syntactic grammars using defined dictionaries. The grammars perform recognition and extraction of the pertinent response according to the extracted keywords from the question.

We present in the following figure the established Arabic syntactic grammars.

The transducer of Fig. 2 describes the different paths allowing the definition of such disease. Each path represents a set of rules extracted from the studied corpus. This transducer shows that the name of the disease is followed by "هو" or "هي".

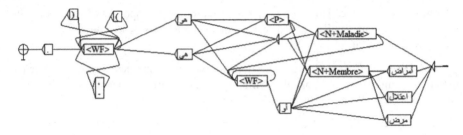

Fig. 2. Transducer recognizing definition of disease

The transducer of Fig. 3 answers the symptoms disease question and causes disease question.

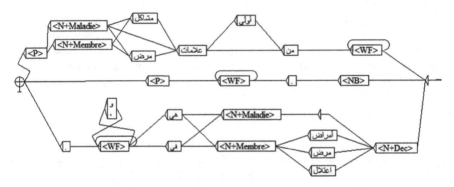

Fig. 3. Transducer recognizing causes or symptoms disease

The transducer in Fig. 4 illustrates the type of disease question belonging to the Arabic corpus.

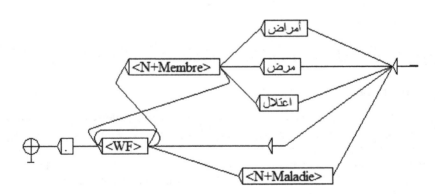

Fig. 4. Transducer recognizing types of disease

It should be noted that Arabic extracted keywords were translated into English. The translated keywords define the appropriate English syntactic grammars that will be applied to English corpora. The English grammars have the same structure of the Arabic ones [7, 8].

4   Experimentation and Evaluation

The experimentation of our question-response system is done with Java. Figure 5 illustrates the prototype interface.

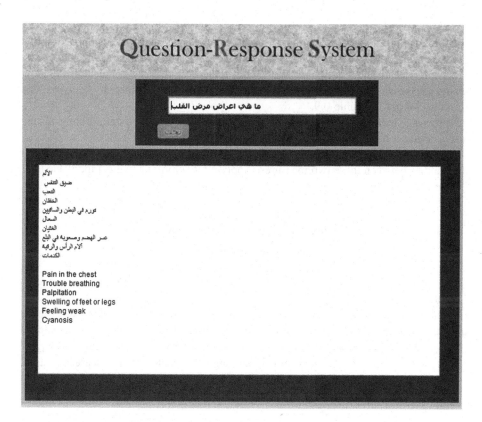

Fig. 5. The question-response system interface

As shown in Fig. 5, the user introduces his question "ماهي اعراض مرض القلب" (What are the symptoms of heart disease?). The system extracts keywords which are "اعراض" (symptoms) and "القلب" (heart). These keywords will be translated into English. According to the extracted keywords, the transducer of Fig. 3 will be applied on Arabic corpora and the corresponding in English will be applied on English corpora using already built dictionaries. Finally, responses are displayed in Arabic and English as illustrated by Fig. 5. It is important to mention that the connection between Java and NooJ is done with noojapply.

To test and evaluate our question-response system, we have collected Arabic and English corpora. Table 4 shows the used corpora.

Table 4. Arabic English corpora

Corpus (Arabic/English)	Number
Definition	200
Symptoms	200
Types	150

We conduct a set of experimentations to evaluate the question-response system's efficiency. In addition to that, we use the following measures for evaluating the results: Precision, Recall and F-measure. We should remember that the recall indicates how many responses among those to be found, are effectively extracted. Applying this formula, we get the value 0.9.

Precision measures the number of relevant answers of the system among all the answers that it has given. Applying this formula, we get the value 0.8.

F-measure is a combination of Precision and Recall for penalizing the very large inequalities between these two measures. Applying this formula, we get the value 0.72 (Table 5).

Table 5. Summarizing the measure values

Precision	Recall	F-measure
0.8	0.9	0.72

The noise is caused by the confusion between the symptoms of disease response and causes disease response in the case of the absence of a trigger. In the sentence يشعر المُصاب بأمراض القلب بألم في الصدر (who has heart disease has chest pain), ألم في الصدر (chest pain) is returned as cause symptom although it refers to a heart disease symptom.

The Arabic and English silence problem is mainly due to the particular structure of some sentences.

5   Conclusion

In this work, we have discussed the problem of access to precise information and more specifically in the medical field. We have focused on question-response systems, based on the NooJ language development platform using a set of lexical and syntactic rules and an elaborate method which also integrates a process of disambiguation that offers us satisfactory results by calculated measures. The purpose here is to generate a precise response to a need for information expressed in Arabic and English.

Scientific research is constantly advancing and as new information emerges others are eliminated, a daily update is very necessary in this field. Another targeted objective is the extension of the coverage of the relations of our medical ontology, i.e. the implementation of the lexico-syntactic patterns method to other relations of ontology, such as the relation: side effects.

References

1. Embarek, M.: Un système de question-réponse dans le domaine médical: le système Esculape. Ph. D. thesis, Université Paris-Est (2008)
2. Fort, K., Ehrmann, M., Nazarenko, A.: Vers une méthodologie d'annotation des entitées nommées en corpus? Traitement Automatique des Langues Naturelles 2009, Senlis, France (2009)
3. Nguyen, T.D., Do, T.T.T.: Natural language question answering model applied to document retrieval system. World Academy of Science, Engineering and Technology 51 (2009)
4. El Ayari, S., Grau, B., Ligozat, A.-L.: REVISE, un outil d'évaluation précise des systèmes de questions-réponses. Actes de Coria (2009)
5. El Ayari, S.: Évaluation transparente de systèmes de questions-réponses: application au focus. RECITAL 2007, Toulouse (2007)
6. Seideh, M.A.F., Fehri, H., Haddar, K.: Named entity recognition from Arabic-French herbalism parallel corpora. In: Okrut, T., Hetsevich, Y., Silberztein, M., Stanislavenka, H. (eds.) NooJ 2015. CCIS, vol. 607, pp. 191–201. Springer, Cham (2016). https://doi.org/10.1007/978-3-319-42471-2_17
7. Fehri, H., Haddar, K., Ben Hamadou, A.: Recognition and translation of Arabic named entities with NooJ using a new representation model. In: FSMNLP 2011. France, pp. 134–142 (2011)
8. Fehri, H., Haddar, K., Hamadou, B.A.: A new representation model for the automatic recognition and translation of Arabic named entities with NooJ. In: RANLP 2011. Hissar, Bulgaria, pp. 545–550 (2011)

NooJ Morphological Grammars
for Stenotype Writing

Mario Monteleone[⊠], Raffaele Guarasci, and Alessandro Maisto

Dipartimento di Scienze Politiche, Sociali e della Comunicazione,
Università degli Studi di Salerno, Salerno, Italy
{mmonteleone, rguarasci, amaisto}@unisa.it

Abstract. Stenotyping is a writing method system used to transcribe spoken texts, rapidly and in real time, using a mechanical or digital device equipped with a special keyboard. This device is called a stenotype, stenotype machine, shorthand machine or steno writer, and it is a specialized chorded keyboard or typewriter allowing to performing beats of one or more keys simultaneously. Stenotyping requires the application of specific coded writing systems intended to limit and accelerate the number of beats. Whereas high-speed beats often generate a high amount of typos, the creation of a stenotype writing method based on a non-casual combination of morphemes would rely on a defined list of elements to be combined (i.e., the morphemes of a language) together with a production syntax (that is, the morphological rules of a language). Therefore, in this paper, we will show how to use NooJ linguistic resources and morphological grammars to build and implement a system for real-time typos automatic correction during stenotype writing.

Keywords: NooJ · NooJ local grammars · Stenotype writing · Stenotyping
Stenograph · Metodo Melani

1 Introduction

Stenotyping [1] is a writing method system used to transcribe spoken texts, rapidly and in real time, using a mechanical or digital device called a stenotype, stenotype machine, shorthand machine or steno writer. It is a specialized chorded keyboard or typewriter allowing to performing beats of one or more keys simultaneously. Stenotyping requires the application of specific coded writing systems intended to limit and accelerate the number of beats[1].

---

[1] Especially in the USA where stenotyping is a legally defined profession. In order to pass the United States Registered Professional Reporter test, a trained court reporter or closed captioner must write speeds of approximately 180, 200, and 225 words per minute (wpm). Some stenographers can reach 300 wpm.

© Springer International Publishing AG 2018
S. Mbarki et al. (Eds.): NooJ 2017, CCIS 811, pp. 200–212, 2018.
https://doi.org/10.1007/978-3-319-73420-0_17

The Stenograph is used with a specially modified version of the "Metodo Melani"[2], one of the most worldwide used systems for stenotyping. However, in the adaptation to computers, this system presents the limit of requiring the use of abbreviations for syllables and/or word portions, without taking into account the morphemic structure of words. Considering that the speed of request beats often generates a high amount of typos, the creation of a stenotype writing method based on a non-casual combination of morphemes would rely on a defined list of elements to be combined (the morphemes of a language) together with a syntax (the morphological rules of a language).

Therefore, in this paper, we will show how to use NooJ linguistic resources and morphological grammars [2–4] to build a system for real-time typos automatic correction during stenotype writing, i.e. a routine for most modern computerized shorthand machines. In order to account for all catalogued words of lexical constellations, our grammars will describe Italian morphological segmentation and morpheme combinations. All instructions in the grammars will be associated to specific chording or stroking patterns, allowing shorthand machines to spell out not syllables or portion of words, but morphemes, sequences of morphemes and complete words. The system will use the Italian NooJ electronic dictionaries to test and validate the produced combinations. The method, described here, can be adapted to any language having a formalized morphology.

2 The Stenograph Machine and the "Metodo Melani"

The keyboard [5] of a stenograph machine (see Fig. 1) has only 23 keys: although graphically split, the letter "S" to the left of the keyboard and the asterisk both are single keys. Multiple keys are pressed simultaneously (a procedure known as "chording" or "stroking"), in order to spell out whole syllables, words, and phrases with a single hand motion. The keyboard does not cover all complete alphabets[3], and

---

[2] In 1980, Professor Marcello Melani, a scholar and teacher of shorthand and computer analysis, conceived the Melani-Stenotype method for the Italian language, directly compatible with electronic processing. The widespread use in the American courts and the forthcoming reform of the Criminal Procedure Code (...) pushed prof. Melani to design a "method in view of the application to the computer for automatic transcription... without the need for subsequent interventions in terms of additions, changes, corrections, except for any errors committed by the operator during registration". In the preface to his manual of 1994, Melani states that "... developments in computer technology had enabled shorthand machine to take advantage of electronic processing for a stenotype automatic transcription, and can therefore realize the mirage of real-time writing, which has now become a reality, not theoretical but practical." The Melani system was born "taking into account the compatibility with the stiffness and potential offered by computers, especially in terms of real-time support. In this respect, the computer input, which is also essential to provide a modern, efficient and automatic stenotyping, is relatively simple: there is no need for sophisticated algorithms or particularly large dictionaries of abbreviations; the computer simply recognizes some shortened codes, conceived from the beginning so as not to yield to ambiguities, and managing a dictionary of acronyms." The main feature of Melani's technical system is a not-abbreviated writing of the text, i.e. the real-time production of verbatim computer-assisted reports. The "Metodo Melani" had a great success, and in the '90s, its author adapted it to Spanish and Portuguese.

[3] Some languages, as French, English and American English, use modified types of keyboards.

Fig. 1. Stenotype keyboard sample

stenotype writers use combinations of keys to substitute missing letters. In Italy, the widespread diffusion of stenotypia took place around the early 90s, when the first machines and methods appeared on the market, addressed to those users needing to transcribe correctly speeches as quickly as possible. During this period, the most relevant Stenograph stenotype machine and the already mentioned "Metodo Melani" became both so widespread that today are generally referred to by the unique expression "Metodo Melani". To achieve writing, the "Metodo Melani" disassembles the keyboard into two basic parts: one operated by the left hand, one operated by the right hand.

Fig. 2. The left part of a Stenograph keyboard.

Stenotype writers use the left part of the keyboard to write consonants that occur as first letters in words or syllables. These consonants are classified as word or syllable initials. For instance, the letter "p" in the Italian preposition "**per**" (for) will be a word initial, while the same letter in the word "**pe**-san-te" (heavy) will be a syllable initial. To write word/syllable initials, left-hand fingers will strike the keyboard in succession, from left to right, i.e.: little finger, ring finger, middle finger, and index finger.

Fig. 3. The middle-lower part of a Stenograph keyboard.

Stenotype writers use this part of the keyboard to write vowels that occur in the initial part of syllables or in the middle part of words. These vowels are classified as syllable/word initial or middle letters. For instance, the vowel "**e**" in the word "**e**ra" (era) will be a syllable/word initial letter; the same vowel in the word "ben**e**" (good) will be a middle word letter. To write syllable/word initial or middle letters, fingers will strike the keyboard in succession, from left to right, i.e.: the left-hand thumb finger for the vowels "i" and "a", the right-hand thumb finger for the letters "e" and "o". Instead, the vowel "u", not displayed in this part of the keyboard, will be written stroking contemporarily the keys "e" and "o" with the right-hand thumb finger.

Fig. 4. The central-right part of a Stenograph keyboard.

Stenotype writers use this part of the keyboard to write consonants which occur in the middle or final parts of words. These vowels are classified as median or final word consonants. For instance, the consonant "r" of the word "ara" (are) will be a word median consonant; instead, the same letter in the word "bar" will be a word final consonant. To write median or final word consonants, right-hand fingers will strike the keyboard in succession, from left to right, i.e.: index finger, middle finger, and ring finger.

Fig. 5. The right part of a Stenograph keyboard.

Stenotype writers use this part of the keyboard to write vowels occurring at the end of words. These vowels will be defined as final-word. They are typed exclusively at the end of a word, even if the word itself is composed by only one syllable. When one of these letters is typed, and the machine is connected to the computer, a space bar between the written word and the next one is automatically entered. The only finger used to type word vowels is the little finger of the right hand. Instead, the vowel "u", not displayed in this part of the keyboard, is written stroking contemporarily the keys "a" and "o", using the right-hand little finger. As for the letters not present on the Stenograph keyboard, it is necessary to use key combinations. For instance, considering the left part of Fig. 2, the consonants occurring at the beginning of words or syllable are written by combining contemporarily the following keys: PTV = **B**; TH = **D**; TV = **F**; PC = **G**; HR = **L**; CHR = **M**; H = **N**; PTVCHR = **Q** (six keys simultaneously pressed exactly on the splitting slot); SPT = Z.

As already stated for the middle-lower part of the keyboard in Fig. 3, the vowel "u" is written stroking contemporarily the keys "e" and "o" with the right-hand thumb finger.

As for the central-right part of the keyboard in Fig. 4, the median or final word consonants not present are formed stroking contemporarily the following keys: CTP = **B**; TH = **D**; TP = **F**; PR = **G**; HR = **L**; SHR = **M**; H = **N**; CTPSHR = **Q** (six keys simultaneously pressed exactly on the splitting slot); CT = **V**; SH = **Z**.

As already stated, the vowel "u", not displayed in the right part of the keyboard in Fig. 5, is written stroking contemporarily the keys "a" and "o" by means of the right-hand little finger. The "Metodo Melani" allows the encoding of the most frequently used words, and subsequently their association to the stroking of specific keys on the keyboard. For instance, to write a word such as "circostanziale" (circumstantial) we will strike contemporarily the keys CSO. As it is visible in Fig. 1, the Stenograph keyboard is also equipped with an autonomous bar of numbers. Numbers are typed as follows: left part of the bar/keyboard: "S + bar" for number "1", "P + bar" for number "2", "T + bar" for number "3", "V + bar" for number "4", "I + bar" for number "5"; the right part of the bar/keyboard: "C + bar" for number "6", "T + bar" for number "7", "P + bar" for number "8", "I + bar" for number "9". As for the key with the asterisk, it is used both to write the asterisk itself, when stroked contemporarily with the middle vowel "a", and to delete the last written word, when the stenotype machine is connected to a computer.

3   The Limits of the "Metodo Melani"

The "Metodo Melani" is a completely mnemonic procedure forcing stenotype writers to remember the keys to stroke on the base of the positions the letters have inside words. Therefore, the whole writing process presents a high percentage of errors, also because of the high typing speed. As for word subdivision, this method is not based on the identification and coding of morphemes, but at best of syllables.

This is a crucial computational limit, because the subdivision of words into syllables does not return fixed minimum units. Actually, no specific lists of syllables exist *a priori* in any given language, and the syllable forms and contents may vary according to the words hyphenated. Conversely, the subdivision of words into morphemes restores fixed, stable, and reusable minimum units, definable thanks to the specific word-formation rules pertaining to each language, ranging from inflectional to polysynthetic ones.

The "Metodo Melani" cannot be considered a standard in computational/combinatory procedures, considering that the instructions it uses are not taxonomically coded on the base of their functions, and are not iteratively reusable. Due to these limits, any eventual automatic correction of typos is possible only during post-editing, i.e. at the end of the word typing. At any rate, before going further in our analysis, we want to clarify that we do not want to state the "Metodo Melani" unsuitable for the purposes it was created for, and which it aims at achieving. It has long been the best possible solution for stenotype writers, and still is. However, in our computational-linguistic optics, we are confident that a different, more linguistically-natural stenotypic writing method can be created anew and used to reduce the percentage of errors, providing at the same time automatic correction tools during typing.

4 A Brief Sketch on Italian Morphology

Italian simple-words formation combines lexical, inflectional/grammatical and derivative morphemes, which are minimal units, or immediate constituents, identifiable by word segmentation, and never altered by the contexts in which they occur: each alteration of a morpheme due to its co-occurrence context automatically creates another morpheme. Any word must include at least one lexical morpheme.

Word morpho-semantic features rule the additional presence of inflectional/ grammatical morphemes, in the classical Saussurian terms of "langue" and "parole", as well as of "signifier" and "signified". Inflectional words, even derived ones, always end with an inflectional/grammatical morpheme. Finally, derived non-inflectional words always end with a derivate morpheme.

As many other lexicons, the Italian one is often subject to changes and modifications: new words need to be entered; others fall into obsolescence. However, the morphological features of new words are largely predictable: their "creation" takes place mostly on an analog and iterative basis, in compliance with the already existing norms of use[4].

Besides, although the lexicon of a language is an open set, on the contrary the morphemes it uses form a more stable and closed set, with the exception of lexical morphemes. Thus, combinations of morphemes can be reduced to algorithmic descriptions, likely to be formalized and factorized. Factorization, as we shall see, will be of utmost importance for the realization of the new stenotypic writing method we are proposing to achieve.

5 Italian Morphology and Stenotype Writing

Let us consider the following list of Italian morphemes:

1. a	4. azion	11. he	18. it
2. à	5. e	12. hi	19. ital
3. abil	6. es	13. i	20. mente
4. angl	7. fobi	14. ian	21. o
1. a	8. foni	15. ic	
2. ar	9. franc	16. ism	
3. at	10. grec	17. istic	

---

[4] For example, "americanizzare" (to americanize) has the same meaning model of verbs in which the derivative suffix –izz adds the sense of "making something equal to". Together with the sequence of morphemes americ + an, it acquires the meaning of "making someone or something American". The same formation and structure is present in "finlandizzare" (to finlanidize), "latinizzare" (to latinize), or "talebanizzare" (to talibanize).

A lexicon-based combination of these morphemes gives the possibility to form and transcribe correctly the following 130 Italian words:

1. anglicana
2. anglicanamente
3. anglicane
4. anglicani
5. anglicanismi
6. anglicanismo
7. anglicano
8. anglicismi
9. anglicismo
10. anglicistica
11. anglicisticamente
12. anglicistiche
13. anglicistici
14. anglicistico
15. anglicizzabile
16. anglicizzabili
17. anglicizzabilità
18. anglicizzare
19. anglicizzata
20. anglicizzate
21. anglicizzate
22. anglicizzato
23. anglicizzazione
24. anglicizzazioni
25. anglismi
26. anglismo
27. anglo
28. anglofoba
29. anglofobe
30. anglofobi
31. anglofobia
32. anglofobo
33. anglofona
34. anglofone

35. anglofoni
36. anglofonia
37. anglofono
38. franca
39. francesismi
40. francesismo
41. francesistica
42. francesisticamente
43. francesistiche
44. francesistici
45. francesistico
46. francesizzabile
47. francesizzabili
48. francesizzabilità
49. francesizzare
50. francesizzata
51. francesizzate
52. francesizzate
53. francesizzato
54. francesizzazione
55. francesizzazioni
56. franche
57. franchi
58. franco
59. francofoba
60. francofobe
61. francofobi
62. francofobia
63. francofobo
64. francofona
65. francofone
66. francofoni
67. francofonia
68. francofono

69. greca
70. greche
71. greci
72. grecismi
73. grecismo
74. grecistica
75. grecisticamente
76. grecistiche
77. grecistici
78. grecistico
79. grecizzabile
80. grecizzabili
81. grecizzabilità
82. grecizzare
83. grecizzata
84. grecizzate
85. grecizzate
86. grecizzato
87. grecizzazione
88. grecizzazioni
89. greco
90. grecofoba
91. grecofobe
92. grecofobi
93. grecofobia
94. grecofobo
95. grecofona
96. grecofone
97. grecofoni
98. grecofonia
99. grecofono
100. itala
101. itale
102. itali

103. italianismi
104. italianismo
105. italianistica
106. italianisticamente
107. italianistiche
108. italianistici
109. italianistico
110. italianizzabile
111. italianizzabili
112. italianizzabilità

113. italianizzare
114. italianizzata
115. italianizzate
116. italianizzate
117. italianizzato
118. italianizzazione
119. italianizzazioni
120. italo
121. italofoba
122. italofobe

123. italofobi
124. italofobia
125. italofobo
126. italofona
127. italofone
128. italofoni
129. italofonia
130. italofono

It is worth noting that in this list "anglistica" (Studies on English literature and language) and "italianistica" (Studies on Italian literature and language) are both names and adjectives. Adding only one morpheme, i.e. "german", it increases the productivity of our lexicon-based combination, which reaches the number of 161 attested words:

131. germana	142. germanizzabili	153. germanofobe
132. germane	143. germanizzabilità	154. germanofobi
133. germani	144. germanizzare	155. germanofobia
134. germanismi	145. germanizzata	156. germanofobo
135. germanismo	146. germanizzate	157. germanofona
136. germanistica	147. germanizzate	158. germanofone
137. germanisticamente	148. germanizzato	159. germanofoni
138. germanistiche	149. germanizzazione	160. germanofonia
139. germanistici	150. germanizzazioni	161. germanofono
140. germanistico	151. germano	
141. germanizzabile	152. germanofoba	

6 Morpheme Combinatorial and Stenograph Keyboard

Considering what has been stated so far, our method will be therefore divided into the following steps: a new subdivision of stenograph keyboard on the basis of Italian implicit phonological rules; the detection and classification of all the morphemes of the Italian language; the numeric and/or alpha-numeric tagging of each single morpheme by means of unique and not ambiguous labels; the association of each tag or sequences of tags, to one or more keys of the Stenograph machine, also using keystrokes; the factorization of the typing rules obtained; the use of Italian DELAF (over two million entries) to control automatically the words typed.

This method will imitate the word formation practices typically used by native speakers, thus granting a more natural writing procedure than the one offered by the "Metodo Melani".

Each morpheme will match a single tag that will in turn correspond to a single key or a single sequence of keys of the stenograph machine. Along with the aforementioned recursive word-forming modes, such a unique relationship between morphs, tags, and keystrokes will greatly lower the percentage threshold of typing errors. Despite its limited number of keys, the Stenograph keyboard can provide a very large number of potential combinations, i.e.: $23! = 25.852.016.738.884.976.640.000$. However, keystroke options depend on the maximum number of contemporarily usable fingers, which is equal to 10. Thus, the number of possible combinations is: $23^{10} = 41.426.511.213.649$. If we consider that inflectional languages have a maximum average of 3 morphemes per word, the number of words that can be written with the Stenograph keyboards will be on average: $13.808.837.071.216,33$ (periodic), which is a much larger number than the total of words used in Italian, be it in its function of current, specialized or domain language.

7   NooJ Morphological Grammars for Stenotype Writing

As already mentioned, the missing letters on the Stenograph keyboard are written pressing more keys simultaneously. For example, the letter "b" occurring at the beginning of a word is entered by typing the keys PTV; at the same time, when occurring at the middle or at the end of a word, it is entered by typing the keys CPT.

To overcome the mnemonic nature of these procedures, as for Italian phonology, our approach will use the rule f minimal pairs, which is: Alveolars Fricatives: s → z, Bilabial Occlusives: p → b, Velar Occlusives: c → g, Labiodental Fricatives: v → f, Alveolar Vibrant and Approximant: r → l, Alveolar Occlusives: t → d.

Therefore, a key/letter that transcribes a specific sound (for example, the "b" in bet, which is a voiced bilabial occlusive) can be used also to write a letter that has some similar characteristics (in this case, the "p" of pet, which is an unvoiced velar occlusive). For instance, we may establish that the "P" will always correspond to the stroke sequence PTV, and "B" to the stroke sequence PPV.

In addition, the key H will be associated with the pair of nasal consonants "m" and "n", the vowel "u" will be given by the combination EO; while the consonant "q", always co-occurring with the vowel "u", will be given by the combination of SO. Besides, accented letters will be written by doubling any given vowel. In Italian, letters are accented only in the final position, for specific words that are composed by at least two syllables.

Their accent is always grave, except for preset words (i.e. some conjunctions and verb voices) that require an acute accent. Therefore, their correct identification will come directly from the Italian DELAF. Our approach will not increase the number of letters to stroke, and will follow the criteria of the speaker's intuition and phonological logic.

As for the Italian morphology, in our method, all the keys composing a word (in the sense of a sequence of morphemes) will be stroked simultaneously. Our basic idea is to introduce a level of morphological analysis that allows the typing of the word by segmenting it into its main morphemes. To achieve this segmentation, we will build a specific NooJ dictionary, an example of which is shown below:

#lessicali	ic,SFX+type=ic	es,SFX+type=es
ahcr,LEX+type=angl	hht,SFX+type=mente	iah,SFX+type=ian
vrahc,LEX+type=franc	iss,SFX+type=izz	
cerhah,LEX+type=german	apr,SFX+type=abil	#suffissi grammaticali
crec,LEX+type=grec	itaa,SFX+type=ità	a,GRM+type=a
itar,LEX+type=ital	are,SFX+type=are	e,GRM+type=e
	at,SFX+type=at	i,GRM+type=i
#suffissi	ash,SFX+type=azion	o,GRM+type=o
ish,SFX+type=ism	ovp,SFX+type=ofob	he,GRM+type=he
ist,SFX+type=ist	ovh,SFX+type=ofon	hi,GRM+type=hi

Morphemes are set into three groups (lexical, inflectional/grammatical and derivative ones). The entries of this dictionary are sequences of stenograph keys, to be stroked contemporarily. By means of the instructions in this dictionary, such sequences are put into correspondence (i.e. are rewritten as) specific Italian morphemes. Finally,

NooJ morphological grammars will describe the needed word segmentations, and check for the presence/absence of typos, and suggest revisions/corrections.

For instance, let us consider the word "germanofobo" (germanophobe). First, it will be segmented into its main morphemes, which are: one lexical morpheme "german", two grammatical morphemes "o", and one derivative morpheme "fob". By factorization, the first grammatical morpheme "o" will be agglutinated to the morpheme "fob". Therefore, the "ofob" sequence will be written by typing the sequence *OVP, omitting the central vowel.

An example of this procedure is displayed in the following graph, which shows the segmentation of the noun "italianistica" (Italian studies) plus the two adverbs "italianamente" (in the Italian way) and "italianisticamente" (in the Italianistic way). Any different sequence of letters not complying with those in the graph will not come to the end of it, being therefore a typo (Fig. 6):

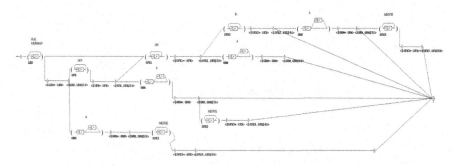

Fig. 6. Morphological segmentation of the words, "germanamente", "germanistica", "germanisticamente" "italianamente", "italianistica", "italianisticamente".

The previous grammar, together with the already mentioned dictionary, will process and annotate a text, producing the following results (Fig. 7):

The rewriting of words will then take place in different steps: morphemes will be typed by means of their corresponding strokes, and follow their specific sequences. For example, with regard to the word "germanizzabilità" (germanizability"), the lexical morpheme will be written stroking contemporarily the keys CERHAH (in the following scheme, the capital bold letters correspond to simultaneously pressed keys) (Fig. 8):

s p t v * c t p i **A**
s **C H** r * s **H R** e o
i a **E** o

Then, the additional successive morphemes will be written in sequence:

ISS	APR	ITAA
S p t v * c t p i a	s p t v * c t **P** i a	s p t v * c **T** p i **A**
S c h r * **S** h r e o	s c h r * s h **R** e o	s c h r * s h r e o
I a e o	i **A** e o	**I A** e o

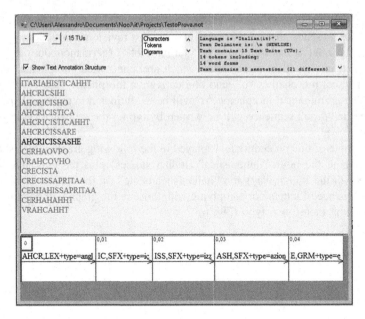

Fig. 7. Example of stenograph annotations in a text

Fig. 8. NooJ dictionary of the annotated text

Besides, syntactic grammars (i.e. transducers, as the one of Fig. 9) will use the results of the annotation procedure to rewrite Stenograph key sequences into Italian words. The numbered variables are to be considered as slots, which may be full or empty, depending on the morphological complexity of the words accounted for:

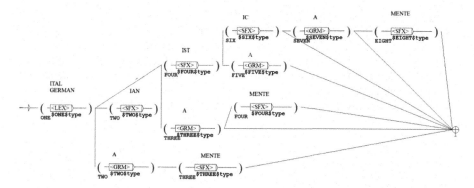

Fig. 9. NooJ FST for the transformation of keystrokes into Italian words.

NooJ concordances can be used to control the written text:

Finally, all "Unknown Words" will be considered as either errors or new dictionary entries (Fig. 10).

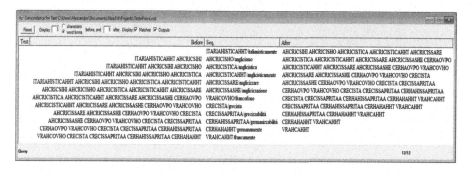

Fig. 10. NooJ stenotype concordances

8 Conclusion and Future Works

To complete our method, we will achieve the following tasks: the creation of an exhaustive Italian morpheme dictionary for NooJ, the association of each morpheme to specific keystrokes on the Stenograph keyboard, the factorization of such associations, and the creation of NooJ grammars for word segmentations and typo correction. All these linguistic resources will become part of a Python Module, so to be used as both standalone routines and NooJ extensions. It will also be possible to build a unique Python Module appropriate to include the NooJ environment.

References

1. https://en.wikipedia.org/wiki/Stenotype
2. Silberztein, M.: The NooJ Manual (2003). www.noj4nlp.net
3. Silberztein, M.: La formalisation des langues: l'approche de NooJ. ISTE Ed.: Londres (2015)
4. Silberztein, M.: NooJ computational devices. In: Donabédian, A., Khurshudian, V., Silberztein, M. (eds.) Formalising Natural Languages with NooJ, pp. 1–13. Cambridge Scholars Publishing, Newcastle (2013)
5. http://www.accademia-aliprandi.it/repertori/doc/Macchine_stenotipiche_italiane.doc

From Language to Social Perception
of Immigration

Carmela Scoppetta[(✉)], Anastasia Alfieri, Flavio Merenda, Sonia Lay,
Annalisa Colasanto, and Raffaele Manna

Dipartimento di Scienze Sociali Politiche e della Comunicazione,
Università di Salerno, Fisciano, Italy
scoppettacarmen@gmail.com, flavio.mrn@gmail.com,
sonia.lay84@gmail.com, colasantoannalisa@gmail.com,
raffaele.manna@gmail.com

Abstract. Mass media plays a very important role in shaping and controlling
public opinion. This study focuses on exploring how media affects Italian
people's attitudes and perceptions towards one of the most complex social
phenomena of modern society, immigration. The aim of the authors is to be both
comprehensive and neutral in their analysis. As stated in the Cambridge Dictionary, "Immigration is the act of someone coming to live in a different
country", and involves "the process of examining your passport and other
documents to make certain that you can be allowed to enter the country, or the
place where this is done" (http://dictionary.cambridge.org/dictionary/english/
immigration). Data was collected from different sources: a corpus of 180 articles, taken from blogs and daily newspapers of different political orientations,
and a second corpus including people's comments related to these articles.
NooJ's multi-layer approach has allowed the research group to create different
syntactic grammars and domain-dictionaries, with the purpose of mining relevant pieces of information from the corpora and comparing the results collected
by each member. The group has performed several analyses to assess the
validity of preliminary hypotheses on "How media convey dominant ideologies"
and, consequently, "How Italian people perceive immigration". In doing so, it
was possible to extract and compare data providing information about the main
messages spread to the population, as well as other significant aspects of the
analyzed phenomenon.

Keywords: The social perception of immigration · Media
Dominant ideologies

1 The Creation of a Domain Dictionary on the Theme
of Immigration in Italy

In the process of creating the domain dictionary, we have started with the tokenization
of the whole corpus, to identify a first terminological nucleus to include in the dictionary. First of all, we have given priority to the most frequent terms within our
corpus. We have assigned a different tag to the forms, according to their grammatical
category. This operation, known as part of speech tagging, is fundamental to the

© Springer International Publishing AG 2018
S. Mbarki et al. (Eds.): NooJ 2017, CCIS 811, pp. 213–224, 2018.
https://doi.org/10.1007/978-3-319-73420-0_18

procedure of disambiguation of the words. We have defined DELAS or electronic Dictionary of the simple forms, as the set of the canonical forms of a language opportunely labelled. Subsequently, we have created a grammar including the codes to be applied to the different voices of our dictionary. After applying the automatic inflection, we have obtained the DELAF, or the electronic dictionary of the inflected forms. For instance, we will focus on the word <immigrant> that may be considered either as an adjective or a noun.

Analyzing the paradigm, we get immigrato-a-i-e. For this type of inflection, we have inserted the code F1. In the creation of the dictionary, we define, therefore, the grammatical category of the word followed by the code of inflection.

immigrato, N + FLX = F1
but also
immigrato, A + FLX = F1

The analyzed corpus contains 17365 tokens. We have selected only the most relevant ones to the aim of our research; we have obtained a terminological domain dictionary containing 141 lexical forms for a total of 317 inflected forms

2 The Voice of Italy About Immigration

2.1 Introduction

This part of the analysis has led to the creation of a syntactic grammar, with the aim of extracting Entity Names characterized by the presence of the noun "*immigrato*" (in English "*immigrant*") as the head of the sentence, its flexive paradigm and synonims. Specifically, this section has focused on adjectives, past participles and relative clauses. These parts of speech modify the head of the Entity Name.

2.2 The Terms Underlying the Concept of Immigration

Nooj allows working with concepts through a semantic expansion process. This part of the analysis has led to the creation of a syntactic grammar about immigration.

Starting from the word immigrant, the relationships between the terms were investigated considering their synonymy, hyperonymy and hyponymy bonds. The grammar has produced 1239 results. Through the specific module provided in Nooj, it was also possible to perform statistical analyses on the concepts, which have clearly shown the following results:

– "Migrante" is undoubtedly the most frequent term, followed by "immigrati";
– Other terms, like <refugee>, <sheltered>, <migrant> and <immigrant> are often used as synonyms to describe different situations. Actually, our analysis clearly shows the importance of distinguishing between the terms <immigrato> and <migrante>, which are often used interchangeably without recognizing their specificities. The person who migrates is generally the one who moves, and he/she becomes immigrant only after taking the decision to settle in a country that is different from the native land. The clandestine is an irregular migrant; the refugee is the person

who abandons his/her home country to find a shelter; eventually, <profugo> is the one who runs away from war, poverty, hunger and other natural calamities. The main difference between a "rifugiato" and a "<profugo>" (both terms often translated as "refugee" in English) is that the status of "rifugiato" is recognised by the law.

2.3   The Syntactic Grammar

In order to extract Entity Names concerning *"immigrato"*, the following lemmas, obtained with Token and Digram components, were chosen as a list of synonyms: *migrante, immigrato, straniero, profugo, rifugiato, richiedente asilo.* This list of lemmas was considered to be the head of the Entity Name. The syntactic grammar, which aims at extracting adjectives and relative clauses, was finally built, and the linguistic unit *"campo profughi"* was excluded, because we were searching for human entities.

2.4   Results

The linguistic pieces of information extracted and sorted out were 8366. Therefore, a matrix with the results obtained was created, focusing on the following attributes: Keyword, Adjective 1, Adjective 2, Past Participle, Noun, Relative Clauses, Query and Frequency. The following graph shows the adjectives and their frequency:

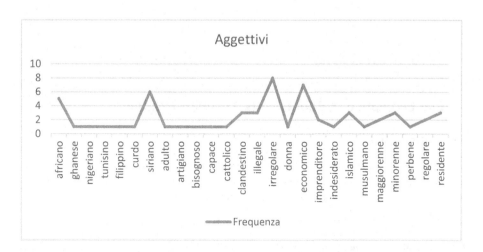

2.5   Semantic Expansion

The second part of this section was dedicated to the creation of a grammar in which the results obtained in the first part were included. The linguistic data extracted through the previous syntactic grammar was labelled in the following way:

1. Perbene (Respectable) – Absolutely positive
2. Bisognosi (Poor) – Positive
3. Clandestini (Illegal immigrants) – Negative
4. Indesiderati (Undesidered) – Absolutely Negative.

2.6   Results II

The results show a similar consideration for the Respectable, Poor and Undesiderable classes. Conversely, the results show a higher frequency with regard to illegal immigration. This means that Italian people usually consider the immigrant as an irregular person, a criminal or a delinquent. The following graph shows the number of times each adjective appears.

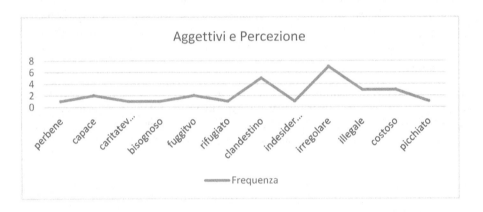

3   The Perception of Immigrants Through Reading

3.1   Introduction

Undoubtedly, the journalistic prose shows fixed politically correct structures and lexicon. These two requisites make it easy to predict syntactic regular structures. Thus, this section focuses on disambiguating some predictable sentences or phrases that occur in the context of news about immigrants.

3.2   Search of Texts

News articles were taken from two national newspapers: LaRepubblica and Huffington Post. Hereafter, this core was enlarged with other news' articles chosen among the blogs of three Italian columnists: Beppe Severgnini, Costanza Miriano and Ernesto Galli della Loggia. Obviously, these news' articles were pre-processed in order to normalize and eliminate undesired data and manage the entire process of study in an appropriate way.

3.3   First Hypothesis

After the corpus was built and the linguistic analysis was done, the list of tokens and bigrams was investigated; they were ordered according to their frequencies, from the higher to the lower ones.

The next step was to run the "Locate" function to look lexemes up and observe their context. All words cover the semantic areas of:

1. The movement of great masses of people in Europe.
2. The documents requested by the bureaucratic system and other related issues.

Examples of this search could be (all words are written in Italian):

– Flusso, movimento, spostamento, migrazione, esodo, etc.
– Richiedere asilo, garanzia, protezione, visto, soggiorno, casa, etc.

Once the results were checked, another simple research was done. The latter aimed at extracting words that were considered relevant because they are often associated with some statistical estimations in the news' reports.

Some examples could be many semi-frozen expressions occurring in Italian language:

– Tasso, aumento del tasso, diminuire, diminuzione, calo, calo del tasso, rischio, crescita, etc.
– Accordo con...
– Proteggere da...
– Difendere da ...

Many results were obtained, so it was necessary to build some syntactic grammars.

3.4   Building Syntactic Grammars

As it is stated above, a disambiguation task was carried out on some frequent phrases, expressions and words. For this purpose, the graphical editor for syntax was used.

First of all, the following word forms were disambiguated:

– Visto: V (past participle) or N (concrete noun)
– Movimento: Concrete or Politics' Noun

The second grammar about *Movimento* as a Concrete or Politics' noun was built. Particularly, this one shows, detects and disambiguates:

– Movements related to the immigrants
– Political parties
– Politically motivated movements
– Terrorist movements

An additional grammar was made to detect and investigate *movimento* and other related forms. It's a little bit more complex because it attempts to detect more structures.

As it is shown below, this grammar is composed by meta-nodes:

- *Fin:* final sentence started with *per*
- *Loc:* locative sentence
- *Pro:* in this one, there are two typical structures with great probability of co-occurrence in that context. That is: *<costringere>* or *<tenere> in schiavitù;* | *<protetta>* or *<difesa> dai* etc.
- *VP:* Verb Phrase node contains verbs and expressions like: *<aumentare>*, *<salire>*, *<crescere>*, *si contrae*.

This grammar is able to detect many syntactic structures used by the reporters.

3.5   Results

Basically, this part of our research did not simply focus on the search for strong and rude words referring to immigrants; our aim was to detect the specificities in writing on this topic, to better understand how this subject is debated through Italian newspapers. Obviously, this study doesn't claim to be exhaustive, but it could be considered as a start for better understanding the impact of the immigrants on the public opinion, the press and people. Actually, it is commonly known that news and media have the power to influence people's attitudes. Language is ambiguous, and many people get easily impressed. In this part of the analysis, nothing was found that we had not expected.

4   The Female Face of Migration

4.1   Introduction and Analysis

The migratory phenomenon has been considered a mainly masculine process for a long time. In the last decades, it has been observed, as Stephen Castles and Mark Miller affirmed that a "feminization of the migratory flows" can be considered as the principal tendencies of the "new era of the migrations". Through the automatic linguistic analysis of our corpus we have tried to bring out the specificities of the female image of migration. Who are the women that decide to abandon their country? What is their nationality? What are the reasons that force them to emigrate? First of all, we engaged in the construction of a syntactic grammar that could answer our questions.

A grammar in the form of a finite state automata that provides 5 different paths has been realized. Each path adds attributes to the central node constituted by the words <woman> and <female>. The meta-node F1 contains another graph in which a relative proposition, connected to the word <woman>, is introduced. The proposition opens with the relative pronoun followed by a verb. At this level, two paths are possible: the verb can be followed either by a nominal syntagm, or by a preposition followed by a noun. Applying the grammar to the text, 201 occurrences were found, of which we analyzed the related linguistic context. We performed an accurate selection of the concordances obtained to remove duplications.

4.2   Results

Women belonging to some different cultural systems are mainly negatively described ("women covered as an armchair to move", or, "the Islamic Republic has a different idea of woman compared to ours", or "women cannot speak, drive or study"). As for the origin, the word woman is often associated with Nigerian, Philippine, Cameroonian or in general African nationalities. Moreover, the word woman is often associated with negative terms like "slave" and "victim". It is the intense traffic of Nigerian women that increases prostitution. But, why do women emigrate towards Italy? Certainly, it is to get away from wars and persecutions. Eventually, we created another grammar that summarizes the obtained results by replacing the labels with more appropriate expressions and words.

5   Tolerance vs. Intolerance

5.1   Introduction

In this part of the analysis, a conceptual development perspective has been adopted. Firstly, the concepts of "Tolerance" and "Intolerance" have been semantically extended, by searching for synonyms with the help of digital tools (like the online virtual thesaurus). Subsequently, the obtained adjectives and nouns have been grouped in three clusters according to their emotional connotation (positive, negative, neutral). Then, the use of regular expressions has made it possible to identify the related occurrences within our corpus.

5.2   The Analysis

The next step was about the development of syntactic grammars aimed at detecting the content words associated with those adjectives and nouns mentioned above; specifically, separate automata have been created for the different clusters of words. Particularly, we searched for those Word Forms that precede or follow the nouns and adjectives included in the nodes of our grammars.

Fear is the predominant negative emotion, occurring 61 times in our corpus. In order to identify the motivations behind such a feeling and the widespread considerations, an automata including the word form "fear" able to extract some longer syntagms, was then created. The obtained concordances revealed that the people's fears might depend on their social perception, can be exploited by populisms, may grow according to the perception of rising crime, and might make it difficult to create a climate of trust in the population.

The words included in the positive cluster occur more frequently in our corpus. "Hospitality" and "integration" show the highest number of occurrences (215 and 88 occurrences, respectively), but they are far from being perceived as a positive solution.

An additional automaton has been created in order to extract some syntagms in which the words "hospitality" and "integration" occur individually or together. The results of the latter have confirmed our previous observations, as proven by the following expressions:

"to give immigrants hospitality is an inevitable process, but it must be managed in a balanced way"; "the false myth of integration at all costs"; "there is no case of peaceful integration between different populations and cultures in the human history, without genocide or invasion"; "the cooperatives gain profits without guaranteeing a decent hospitality"; "the hypocrite left-wing party preserves its capital and interests behind the false myth of welcoming immigrants"; "integration is a complex process".

5.3   Results

The hypotheses on the polarity of the word forms that we initially considered as "positive" have been falsified. The analysis clearly shows the prevalence of negative emotions about immigration, perceived as a threat, which is responsible for increasing the risk of crime and terrorist infiltration.

6   Immigrants and Work: A Phenomenon to be Analyzed

6.1   Introduction

This part of the analysis focuses on a specific aspect of the theme "immigrants" and work.

It is considered appropriate to analyze the relationship between the two topics more closely.

6.2   The Analysis

Grammar – Lavorare.nog

A grammar was created to show how many immigrants had already a job, or at least were searching for it.

The syntactic grammar includes some meta-nodes:

- In GN, a name that can be preceded by a determinant was added:
- In F1, some specific verbs were included, in order to focus on who is already doing or has found a job.

Precisely, this meta-node contains:

- -Verbs preceded by the auxiliary "having"
- verbs indicating a job being found or an action in progress (finding, performing, executing, practicing)

Both nodes are linked to:
Terms of reference (job, occupation, work)

- In F2, a preposition followed by a Noun was inserted.

6.3   Application and Results of the Previous Grammar

Results – Illegal work

Rank	Term	Frequency
1	a lavorare in nero	10
2	per lavorare	10
3	a lavorare a nero	10
4	a trovare lavoro	10
5	chi lavora in agricoltura	4
6	chi lavora in sordina	1
7	per trovare lavoro	1

6.4   Results

The main goal of this part of the analysis is to find out how often and in which way the term "work" is associated with immigration. The results show the actual availability of work for immigrants within the Italian territory; besides, it is extremely relevant to confirm our expectations about how "illegal work" is common.

7   Xenophobic and Racist Verbal Abuse in Web Comments

7.1   Introduction

The main aim of this work is to verify the presence of xenophobic and racist contents associated with web users' comments and to do a diastratic and diaphasic analysis of the verbal abuses.

7.2   Collection of Comments, Building of the Corpus and Row Analysis

The comments were extracted from the same web pages consulted to build the corpus of newspapers. Txt.files were created and sorted out from metadata using Regular Expressions. The different .txt files were stored and used to create a corpus of comments (Commenti_corpus.noc) using NooJ.

Rank	Tokens - corpus of comments	Frequency
1	Immigrati	543
2	Migrant	491
3	Immigrazione	447
4	Clandestine	359
5	profughi	271

Rank	Tokens - corpus of articles	Frequency
1	Migrant	296
2	Immigrati	267
3	Immigrazione	188
4	Stranieri	145
5	profughi	138

The first step was to analyze the numbers of tokens about the main words, to order these ones according to their frequency, and to compare them with the results that came out from the corpus of articles.

7.3   Construction of Syntactic Grammars, Application and Analysis

The text of the comments is closer to the oral language; mistakes and orthographic errors are frequent. For this reason, it was helpful to construct simple and flexible grammars, with few constraints.

1° Grammar. This grammar was built to recognize transitive verb phrases and relative ones. It was considered appropriate to search for those frequent phrases describing the negative actions that immigrants would commit in Italy, such as "stealing", "to commit a crime", "raping" and so on. The central keywords of this work (aimed at building the nominal group associated with the action) were:

- <Immigrato> <Migrante> <Profugo> <Straniero> <Immigrazione> <Migrazione> <Clandestino>

After applying the grammar to our corpus, the query returned 15 results. The data has shown that this kind of phrase is actually a stereotype. It was also possible to notice the debate within opposite political groups. We searched the verbs included in the graph with a Regular Expression:

<rubare>|<delinquere>|<stuprare>|<uccidere>|<portare>|<approfittare>|<distruggere>|<rovinare>|<diffondere>|<violentare>

The Query returned 874 occurrences. This is due to the fact that the subject was long discussed within the articles or in the related comments; the obtained data could be analyzed through different research strategies.

2° Grammar. A second grammar aimed at identifying the names of ethnicities, peoples, nationalities, and religious denominations which are associated with the main keywords of this research, and the related adjectives. The application of the grammar returned 26 occurrences. Through the analysis of the results it was possible to find that the words <africano> and <islamico> are often associated with the words <migrante>, <immigranto> and so on. We applied a Regular Expression built with the found associated names:

<romeno>|<rumeno>|<zingaro>|<cinese>|<bengalese>|<siriano>|<afgano>|<bulgaro>|<turco>|<islamico>|<arabo>|<africano>|<polacco>|<ucraino>|<slavo>|<nero>| <bangladesh>|<indiano>|<pakistano>|<filippino>|<marocchino>|<algerino>|<tunisino>|<colombiano>|<messicano>

The query has confirmed that <africano>, <nero> and <islamico> are the most frequent words within the corpus.

3° Grammar. A third grammar was built in order to find those adjectives that are diastratically and diaphasically placed at a lower level and which are generally associated with violent, xenophobic and racist language. The names associated with these adjectives are the same as the "IMMIGRATI" meta-node of the previous grammar (that expressed members of ethnicities, peoples, nationalities, and religious denominations). The query returned only 12 occurrences. Half of them are really racist and belong to «Libero» and «Il Giornale» web articles. A Regular Expression was built with the «swearing» node («INSULTI»):

<cazzo>|<stronzo>|<coglione>|<merda>|<bestia>|<schifoso>|<puzzolente>|<accattone>|<fottuto>|<scimmia>|<scimpanzé>|<orango>|<pezzo>di<merda>|<mongoloide>|<malato>|<appestato>|<zozzo>|<sudicio>|<sporco>|<bastardo>|<gorilla>| <scimmione>|vu cumpra

It returned 195 occurrences and some of them can be considered verbal abuses. A distributional analysis of the lemmas made it clear that the majority of them belong to «Il Giornale» web articles.

7.4   Results

This part of the analysis has given some expected and conflicting results. The first obstacle was about extracting and predicting the syntactic structures of people's comments, being the texts' error-prone. Some stereotypes emerged. One of the most interesting outcomes was the low presence of racist and xenophobic content, perhaps mediated by some kind of censorship (the phrase "comment removed" was frequent). However, it was not surprising that the lowest diastratic and diaphasic content, often also xenophobic and racist, was more frequent in those articles whose newspapers are ideologically hostile to immigration.

8   General Conclusion

The analysis has highlighted the complexity of a phenomenon that involves different areas of social life and which seems to increase social unrest and fears. The immigrants are defined as: migrants, immigrants, foreigners, refugees. They are considered "people who move from a place to another" causing migratory flows, jobseekers, most of whom women. The comparative analysis has shown that, even if there are not only discriminatory positions towards immigrants, integration is still a problem in Italy. The broadly shared opinion according to which immigration causes an increase in crime is the basis of the dominant feeling of fear. With regard to the origins of immigrants, a lot of people have generally used the term Africans, probably because it is easier to think of immigrants as people that are different from us, thus "black". These reasons make it increasingly difficult to manage migrants' integration problems, thus generating contradictory and often negative opinions about the effectiveness of immigration policies.

References

1. Cheikhrouhou, H.: Recognition of communication verbs with NooJ. In: Koeva, S., Mesfar, S., Silberztein, M. (eds.) Formalising Natural Languages with NooJ 2013: Selected Papers from the NooJ 2013 International Conference (Saarbrucken, Germany), pp. 153–168. Cambridge Scholars Publishing, Newcastle (2014)
2. Ehmann, B., Garami, V.: Narrative psychological content analysis with NooJ: linguistic markers of time experience in self reports. In: Kuti, J., Silberztein, M., Varadi, T. (eds.) Applications of Finite-State Language Processing: Selected Papers from the NooJ 2008 International Conference (Budapest, Hungaria), pp. 186–196. Cambridge Scholars Publishing, Newcastle (2010)
3. Laudanna, A.: Il Linguaggio. Strutture linguistiche e processi cognitivi, Laterza (2006)
4. Pignot, H., Piton, O.: Mary Astell's words in a serious proposal to the ladies (part I), a lexicographic inquiry with NooJ. In: Gavriilidou, Z., Chatzipapa, E., Papadopoulou, L., Silberztein, M. (eds.) Proceedings of the NooJ 2010 International Conference (Komotini, Greece), pp. 232–244. University of Thrace Ed., Greece (2011)
5. Silberztein, M.: NOOJ Manual (2003). disponibile su www.nooj4nlp.net

NooJ's Future

Nooj Graphical User Interfaces Modernization

Zineb Gotti$^{(\boxtimes)}$, Samir Mbarki, Sara Gotti, and Naziha Laaz

Faculty of Science, Ibn Tofail University, Kenitra, Morocco
twinz.gotti@gmail.com, mbarkisamir@hotmail.com,
gotti.sara1990@gmail.com, laaznaziha@gmail.com

Abstract. The legacy of mainframe terminal applications has generally limited the complexity level in desktop applications' user interfaces. Nevertheless the apparition of the new Internet-related technologies is driving to the migration of traditional desktop applications into the web to benefit from the internet technology services. However, GUI's modernization is a new software engineering field that requires a thorough analysis to build and preserve the important characteristics and functionality of the user interfaces. It provides support for transforming existing system's user interfaces to new ones that satisfy new demands. In this work, we have focused on the Architecture-driven modernization ADM approach as a best solution for the legacy system's evolution. The OMG ADM Task Force defines a set of standards to facilitate interoperability between modernization tools. We cite the Knowledge Discovery Metamodel and Abstract Syntax Tree Metamodel. For our work, these two standards will help us to capture design knowledge needed for the construction of modern Nooj user interfaces. We present along this work a reengineering of Nooj application GUIs. We explain its migration process to transform the old desktop GUIs into modern ones respecting web technologies. The process consists of a deep analysis that affects both the structural and behavioral aspects of a GUI, and sophisticated reverse engineering algorithms that must be designed to cope with it.

Keywords: Architecture-driven modernization (ADM) · ADMTF
Reverse engineering · Parsing · Visitor pattern · Graphical user interface (GUI)
Knowledge Discovery Model (KDM)
Abstract Syntax Tree Meta-model (ASTM) · Nooj application
Grammar graphical editor · HTMLM · JavaScriptM

1 Introduction

Generally, software systems often become legacy ones as a consequence of uncontrolled maintenance combined with obsolete technologies. So, the companies must evolve their legacy systems [1]. To evolve such system, we need to identify, understand and adapt the business logic implemented in the source code to facilitate its migration to a new platform.

The Modernization is the practice of understanding and evolving the existing software. It is a process to generate modern systems. In general, it includes all activities related to the improvement of software understanding and various quality parameters, such as the complexity, maintainability, and reusability. Thus, it will extend the

© Springer International Publishing AG 2018
S. Mbarki et al. (Eds.): NooJ 2017, CCIS 811, pp. 227–239, 2018.
https://doi.org/10.1007/978-3-319-73420-0_19

lifetime of a software system. It provides support for transforming an existing software system to a new one like the web that satisfies new demands. Thus, the user can access the application from wherever using the web browser and work with the resources available on the internet, including storage and CPU processing power to help clients with limited hardware capacities. Additionally, the use of web browsers' multimedia capabilities allows users to deal with more interactive and rich user interfaces.

In this article, we propose an ADM based approach allowing the migration of Java Nooj system to the web. To meet this requirement, ADM defines two models [2]: ASTM [3] and KDM [4] that are used to capture design knowledge required to build the future Nooj web version. The ASTM model represents the structural aspect and the KDM specifies a language-independent representation of the programs to be analyzed.

We present below the reengineering of Nooj application GUIs. We explain its migration process that is divided into three phases [5]: Discovery Model, Restructuring, and forward engineering phases, to migrate the old desktop GUIs to modern ones respecting web technologies.

The process consists of a profound static and dynamic analysis of a Nooj GUI to generate models representing all information about the GUI characteristics and functionalities extracted from the applications' source code of Nooj system.

The rest of this paper is organized as follows: Sect. 2 is dedicated to the context of our work. Section 3 presents the modernization case study which is based on the ADM approach. Section 4 covers the related works. Finally, Sect. 5 concludes the work and presents future perspectives.

2 Context

As we have mentioned before, every software system must be updated continuously to take advantage of the new technologies' benefits, control maintenance costs, and preserve complex embedded business rules. However, successive modifications degrade the quality of information system and render it more complex. Moreover, the rapid evolution of technology quickly renders existing technologies as obsolete [6]. Thus, the reengineering has become a solution that allows the reuse of software artifacts by preserving the legacy knowledge of the system to reconstitute it in a new form.

In this work, we opt for the ADM as a concept of reengineering or modernizing existing systems. It counts the standards and basics of model driven development. ADM is an approach that automates the process of extracting the business logic, the GUI's characteristics and functionality of the system. It offers a generic and an abstract solution based on models, allowing the identification, extraction and representation of the business and presentation logic to facilitate its migration to a new platform.

ADM was launched in 2003 by the Object Management Group as an initiative related to building and promoting standards that can be applied to modernize legacy systems. It is the process of understanding and evolving existing software assets for the purpose of software improvement, perfection, and reduction in maintenance effort and cost. It's used also for extending the useful life of the existing applications.

It has emerged as an extension of OMG Model Driven Architecture standard that has been appeared due to the ability of modeling languages in expressing requirements at a high abstraction level [7].

The ADM is the mirror image of MDA; it starts with a software system in order to understand and present all its artifacts as models and establishes model transformations between the different MDA abstraction levels. The basic assumption of models-driven engineering is that models are the correct representation for managing objects within a software engineering process. According to MDA, the models are classified into three types [8]:

- CIM (Computational Independent Model) represents the system requirements. It does not contain any information about the internal structure and other technical details of the system.
- PIM (Platform Independent Model) describes the system details. It is a description of the system structure and behavior, but it does not include any details about the use of the platform.
- PSM (Platform Specific Model) specifies the system implementation in a specific platform.

These models are defined according to four-level architecture as shown in Fig. 1. The M1 level is the Model which is a simplification of a system. The M2 level is the Meta-Model that describes in an abstract way a possible structure of models. A Meta-Model in its turn is described by a Meta-Metamodel defined in the M3 level.

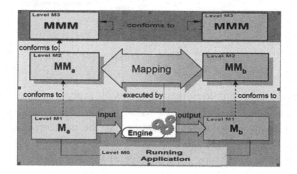

Fig. 1. Manipulation of model level

The manipulation of models or the transition from one model type to another is done via transformations which are some of the important features of model-driven engineering. Two types of model transformations exist: model-to-model transformations and model-to-text transformations. The implementation of model transformations can be carried out using available transformation languages such as QVT language [9] which is proposed as a standard model transformation language by the OMG and which has been used in this work.

As previously mentioned, the ADMTF builds and promotes standards that can be applied to modernize legacy systems. We focus on the KDM and ASTM in particular.

The ASTM standard provides the most granular view of the system architecture. It is predicted for low level modeling and it is close to the source code. It provides an overview of the results of the existing software assets parsing. However, all concepts are from programming language notions which describe how the statements of a software asset are structured.

The ASTM defines a specification for modeling elements and represents them as a tree of AST nodes. Each node is an element of the Java Programming Language. For example, there are nodes for method declarations, variable declaration, assignments, and so on.

While the KDM ISO/IEC 19506 standard is a common intermediate representation for the existing software systems, it represents and manages all aspects of the existing system architecture. The KDM can leverage information captured by the ASTM.

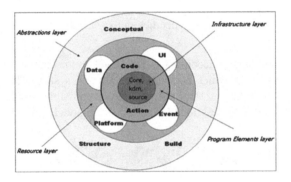

Fig. 2. KDM architecture [4]

KDM is defined as an ontology that describes all aspects of knowledge about the existing application artifacts. The goal of KDM is to ensure interoperability between the tools for maintenance, evolution, evaluation and the modernization of an existing application.

KDM represents the structure, the behavior of the elements as well as the environment that constitutes a system.

This meta-model is used to represent the system artifacts in a high level of abstraction. It provides a knowledge intermediate representation of the existing software systems. Figure 2 shows that KDM has twelve packages organized in four layers. Each package represents software artifacts as entities and relations:

- Infrastructure layer: it includes the Core, kdm, and Source packages. It is the lowest abstraction level; it defines a small set of concepts used systematically throughout the entire KDM specification.
- Program elements Layer: it consists of the Code and Action packages. It represents the code elements and their associations. It consists of a set of meta-model elements common between different programming languages to provide a language-

independent representation. The program elements' layer represents the logical view of a legacy system.

- Resource layer: it represents the higher-level knowledge of the existing software system. This layer focuses on those things that are not extracted from the syntax at the code level but rather from the runtime incremental analysis of the system. There are four packages in this layer: Data, Event, UI and Platform.
- Abstractions layer: it defines a set of meta-model elements for representing domain specific knowledge as well as providing a business overview of legacy information systems. Conceptual, Structure and Build are the three packages in this layer.

3   Modernization Case Study

In this article we opted for a modernization process that transforms an old Java system to the web that satisfies new requirements presented in Fig. 3. The approach is based on the ADM process as a best solution for the legacy system's adaptive and perfective maintenance.

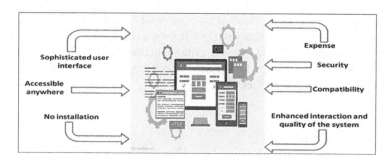

Fig. 3. Web requirements

3.1   The Context of the Case Study

Nooj is a linguistic development software as well as a corpus processor constructed by Max Silberztein and used by linguists [10]. It helps to formalize linguistic phenomena, from spelling to semantics, and to develop orthographical and morphological grammars using either a text or a graph editor.

We will take into consideration the Nooj's graphical editor that provides tools to edit, test and debug local grammars, to apply them to texts. The grammars are represented by organized sets of graphs composed of nodes. Some of them are possibly connected, in which one distinguishes one initial node, and one terminal node. Figure 4 shows the result of Nooj source code analysis which gives an idea on the amount of analyzed data.

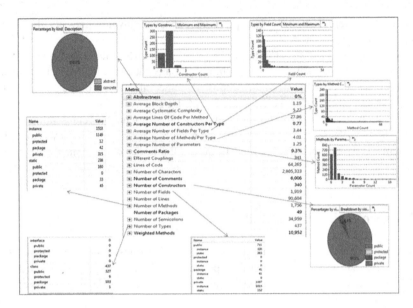

Fig. 4. NooJ source code analysis

The above figure shows the characteristics of Nooj system. For this, we use a Java software testing tool, CodeProAnalytix [11]. It indicates the files' size to give an idea on the amount of analyzed data, as well as the number of the classes treated, which helps us to control and ensure the quality of our approach. All these classes are organized into three packages, (see Fig. 5).

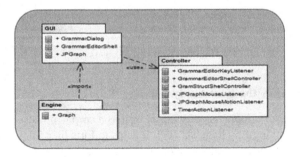

Fig. 5. NooJ platform architecture

Among the most important classes allowing the graphical edition of the grammars, we cite Graph.Java, which is necessary for the graph construction and GrammarEditorShell.Java that instantiates the Grammar.java. Four listeners are analyzed:

- The listener interface for receiving mouse events (press, release, click, enter, and exit) on the grammar editor.

- The MouseMotionListener.Java to track mouse moves and mouse drags.
- The TextBoxKeyListener.Java is a listener of the node's text box.
- The TimerActionListener is a timer that allows the selected nodes flashing.

Our work considers all these features in order to generate modern Nooj GUIs that respect the same appearance of the legacy ones.

3.2 The Proposed Modernization Process

Our contribution considers the ADM approach as a solution to understand and evolve the Nooj system and transform it to a modernized one. It consists of three phases presented in Fig. 6:

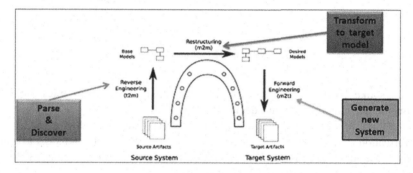

Fig. 6. ADM horseshoe model [2]

- The reverse engineering of the legacy Java Nooj system that represents the extraction of information from the source code at a higher level of abstraction.
- The restructuring and forward engineering: it is a model to model transformations for constructing the target models and generating the new web system from the target models.

The modernization process in Fig. 7 is detailed in the following paragraphs and developed to explain each phase in the process.

For the reverse engineering step of Nooj GUI source code, we use a JDT parser [12] that allows the extraction of Nooj GUI knowledge and business logic. The extracted information will be presented in ASTM [3] and GUIM [13] models. The ASTM model expresses the syntax of the source code, while the GUIM model represents the graphical components, their interrelationships, and their properties; it outlines the containers and the widgets of any main frame as well as their properties.

The parser applies a visitor pattern that allows adding new virtual functions to a family of classes without modifying the classes themselves. It uses the ASTvisitor class to visit all the nodes of the source code. The engine traverses all nodes in an abstract syntax tree and calls the appropriate method (depending on the node type) on each node (ClassVisitor, AnnotationVisitor, FieldVisitor, and MethodVisitor). It uses some

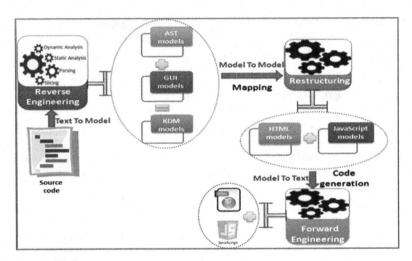

Fig. 7. Overview of the migration process

techniques like parsing and slicing to identify the Nooj business rules and GUI information and ignore irrelevant details.

After that, we transform these two models into KDM model to present and manage all information retrieved at a higher level of abstraction. We develop a model to model transformation in QVTo language. The inputs of this transformation are the ASTM and GUIM models and the output is KDM model.

In this present stage, we opt for a static and dynamic analysis as detailed in [13] to extract any information related to the syntax and structure of the Java source code as well as the presentation of graphical user interfaces and their behavior.

According to ADM, the restructuring and forward engineering steps describe gradual refinement from a higher to a lower level of abstraction. They involve using transformational techniques to automatically obtain the source code according to the new platform or programming language. Once the information retrieved from the user interface is integrated into the KDM model, it can be used and migrated to models with web components. These phases include the web target meta-models definition (javaScriptM) (see Fig. 8) and HtMLM (see Fig. 9), another model to model transformation definition. Finally, the last step is the generation of the target source code in the web platform by keeping the same appearance of the legacy system interfaces. To achieve that, a model to text transformation is defined that takes as input the generated JavaScriptM and HTMLM models. This transformation is based on Acceleo technology, in order to generate the code with web interfaces.

3.3 Obtained Result

We focused on Nooj graphical editor that allows graphical manipulation of grammars. Several analyses have been applied in order to extract all the elements necessary to the edition of the grammars graphically. Here in Fig. 10 there are main interfaces to navigate in order to arrive at the graphical editor.

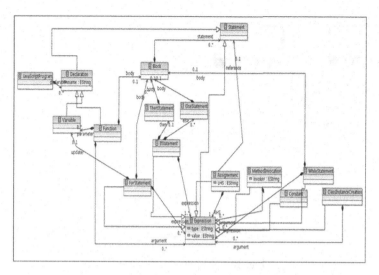

Fig. 8. JavaScriptM Java Script meta-model

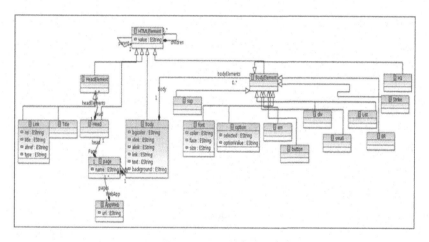

Fig. 9. HTMLM HTML meta-model

The figure below shows the result of the reverse engineering phase presented by ASTM and KDM corresponding to the Node creation case.

As represented in Fig. 11, the execution of a text to model transformation generates a target ASTM model conforming to its meta-model representing the syntax tree of the source code. It is about the structure of the selected source code. The KDM model represents the abstract level of the migration process. It is the result of model to model transformation. Figure 12 presents the responsible algorithm for generating the equivalent KDM model to the node creation.

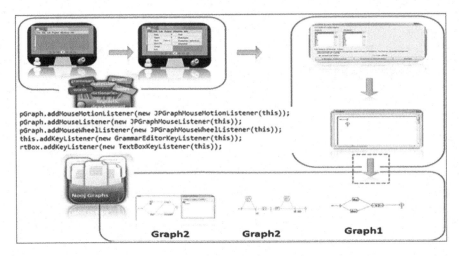

Fig. 10. Nooj graphical editor

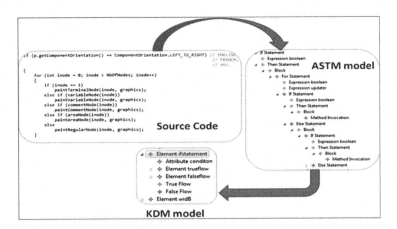

Fig. 11. Reverse engineering result

Figure 13 shows a part of the JavaScript model obtained from the model to model transformation. Figure 14 shows the result of the model to text transformation that generates web interfaces of the Nooj system.

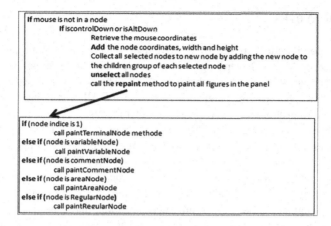

Fig. 12. Node creation algorithm

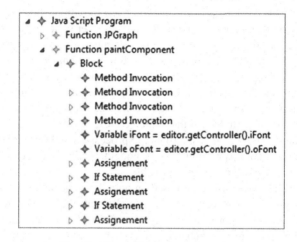

Fig. 13. Example of JavaScript model

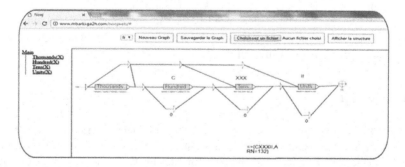

Fig. 14. Nooj web interface

4 Related Works

To obtain a good understanding of system user interfaces, [14] proposed a new approach based on Model Driven Engineering methodology to generate The RIA GUIs structural and dynamic aspects from abstract models. The process is based on Ontology meta-model that presents the logical description of UI components and IFML meta-model that regroups the UI components' interactions.

In [15] the authors, according to MDA approach, used IFML model as the best and abstract solution to model the complex design of user interfaces and interactions. The goal is to execute the IFML due to its executable semantics. The authors introduced an IFVM virtual machine which translates the IFML models into bytecode that will be interpreted later by the Java virtual machine. Other researches in the literature, both on model driven engineering and software modernization, have been presented. The modernization process is based on the analysis of the legacy application code. Two forms of this analysis were introduced: static and dynamic analyses.

Concerning the static analysis, in [16], the authors have applied the static analysis in order to extract the GUI behavioral models from the applications' source code of Java applications. The tool named GUISurfer uses a parser to generate Abstract Syntax Tree (AST) from the application, and a code slicing to extract user interface information, and finally enables the extraction of different behavioral models from the application's source code. Static Reverse Engineering analysis on the source code was performed also by the Modisco tool [17]. It represents an extensible framework to extract information from an existing system. Using its plug-in eclipse, the KDM and UML Model can be generated from the source code.

Moving to the dynamic analysis, a dynamic process named GUI Ripping has been defined. It dynamically builds a running GUI model to facilitate test case creation [18]. It extracts sets of widgets, properties and value.

There are some researches that are based on both the static and dynamic analyses to take advantages and the best features of them. We cite [13], in which the authors proposed an ADM based approach that uses a static and dynamic analysis to obtain knowledge of the structure and behavior of the source code. The approach gives a solution that generates three independent platform combined models (KDM, IFML and TaskModel) for a good understanding and evolving the existing software assets. The models capture various aspects about tasks, presentation and dialog structures and behaviors of the design knowledge, needed for the construction of the future user interface (UI). The authors used a static and dynamic analysis to obtain knowledge of the structure and behavior of the source code.

5 Conclusion

In this work, we have focused on an approach that automates the process of extracting and modernizing the Nooj GUI's characteristics and functionalities. The approach depends on the ADM initiative as the best solution for the legacy system's evolution and it is based on a static and dynamic analysis to obtain knowledge of the structure and behavior of the source code. First, we started by presenting the extracted information in two models which are ASTM and GUIM models, then we transformed these two models into a higher level of abstraction presented by KDM model. After that, in the restructuring and forward engineering phases, the abstract KDM model obtained was migrated into new specific platforms which are the JavaScript and HTML to benefit from the web technologies offered by them.

This automatic process reproduces user interfaces with a modern representation and retains the data related to graphical components namely properties, position and behavior.

In this work we focused only on the NooJ graphical editor. This work can be extended to deal with other NooJ functionalities to be migrated in different platforms.

References

1. Pérez-Castillo, R., de Guzmán, I.G.R., Piattini, M.: Architecture-driven modernization. In: Modern Software Engineering Concepts and Practices: Advanced Approaches, p. 75. IGI Global, Hershey (2010)
2. OMG: Architecture-Driven Modernization. http://adm.omg.org
3. OMG: Abstract Syntax Tree Metamodel. http://www.omg.org/spec/ASTM/
4. OMG: Architecture-Driven Modernization: Knowledge Discovery Meta-Model, v1.4. http://www.omg.org/spec/KDM/1.4/
5. Chikofsky, E.J.: Reverse engineering and design recovery: a taxonomy. J. Softw. IEEE 7(1), 13–17 (1990)
6. Pérez-Castillo, R., de Guzman, I.G.R., Piattini, M., Ebert, C.: Reengineering technologies. IEEE Softw. 28(6), 13–17 (2011)
7. Miller, J., Mukerji, J.: MDA Guide Version 1.0.1. Object Management Group, Needham (2003)
8. Blanc, X., Salvatori, O.: MDA en action: Ingénierie logicielle guidée par les modèles. Editions Eyrolles, paris (2011)
9. OMG: QVT. Meta Object Facility 2.0, Query/View/Transformation Specification. http://www.omg.org/spec/QVT/1.0/PDF/. Accessed June 2014
10. Silberztein, M.: NooJ's dictionaries. Proc. LTC 5, 291–295 (2005)
11. CodePro Analytix. https://developers.google.com/java-dev-tools/codepro/doc
12. JDT: Eclipse Java development tools. https://eclipse.org/jdt
13. Gotti, Z., Mbarki, S.: Java swing modernization approach-complete abstract representation based on static and dynamic analysis. In: ICSOFT-EA, pp. 210–219 (2016)
14. Laaz, N., Mbarki, S.: A model-driven approach for generating RIA interfaces using IFML and ontologies. In: Information Science and Technology (CiSt) (2016)
15. Gotti, S., Mbarki, S.: Toward IFVM virtual machine: a model driven IFML interpretation. In: ICSOFT-EA, pp. 220–225 (2016)
16. Silva, J.C., Silva, C.E., Campos, J.C., Saraiva, J.A.: GUI behavior from source code analysis. In: Interacç ao 2010, Quarta Conferência Nacional em Interacçao Humano-Computador, Universidade de Aveiro, October 2010
17. Eclipse: MoDisco. http://www.eclipse.org/MoDisco
18. Memon, A.M., Banerjee, I., Nagarajan, A.: GUI ripping: reverse engineering of graphical user interfaces for testing. In: WCRE, vol. 3, p. 260, November 2003

A New Linguistic Engine for NooJ: Parsing Context-Sensitive Grammars with Finite-State Machines

Max Silberztein$^{(\boxtimes)}$

Université de Franche-Comté, Besançon, France
max.silberztein@univ-fcomte.fr

Abstract. NooJ is a linguistic development environment that allows linguists to construct large linguistic resources of the four types in the Chomsky hierarchy. NooJ uses a bottom-up, "cascade" approach to sequentially apply these linguistic resources: each parsing operation accesses a Text Annotation Structure, and enriches it by adding or removing linguistic annotations to it. We discuss the drawbacks of this approach, and we present a new approach that requires that all NooJ linguistic resources be represented by a single type of finite-state machine. In order to do that, we must solve theoretical problems such as "how to handle Context-Sensitive Grammars with finite-state machines", as well as some engineering problems such as "how to compose sets of large dictionaries and grammars into a single finite-state machine". Our first experiments show that although that composing large finite-state machines is extremely costly theoretically, the fact that linguistic resources in a typical NooJ cascade depend on each other heavily keeps the size of all intermediary machines manageable. Once the final resulting finite-state machine has been compiled and loaded in memory (e.g. on a webserver) it can be used to parse large texts in linear time.

Keywords: NooJ · RA · Recursive automata · Linguistic engine

1 Introduction

Formal Grammars have been introduced to linguists by (Chomsky 1956) as mathematical tools to describe languages. A Formal Grammar is a set of production rules such as $\alpha \rightarrow \beta$ where both α and β are sequences of empty strings (noted ε), terminal and non-terminal symbols. Depending on constraints on the nature of α and β, (Chomsky 1956) classifies Formal Grammars in four increasingly powerful types: Regular, Context-Free, Context-Sensitive and Unrestricted Grammars. As the power of grammars is augmented, the efficiency of the corresponding parsers is degraded:

© Springer International Publishing AG 2018
S. Mbarki et al. (Eds.): NooJ 2017, CCIS 811, pp. 240–250, 2018.
https://doi.org/10.1007/978-3-319-73420-0_20

parsers for RGs can run in linear time,[1] parsers for CFGs (e.g. CYK) can run in cubic time,[2] parsers for CSGs run in exponential time,[3] and the halting problem of parsers for UGs is undecidable.[4]

Linguists have designed formal notations — i.e. formalisms — to help construct these four types of grammars. For instance, XFST and its variants HFST and SFST[5] are well adapted to the design of RGs; it is straightforward to write CFGs with GPSG;[6] LFG[7] allows linguists to construct CSGs, and HPSG can handle UGs.

Ideally, a linguist would pick each type of formalism according to each type of linguistic phenomenon: for instance, one could use RGs to describe spelling variants, CFGs to compute the structure of sentences and CSGs to describe agreement and distributional constraints. Unfortunately, the aforementioned formalisms are not compatible, making it impossible to include an XFST grammar and a GPSG grammar inside a LFG grammar, or even merge a TAG grammar with a CCG grammar.

NooJ[8] was developed for this reason: it provides linguists with a unified formalism to handle the four types of grammars, thus ensuring that all linguistic resources are compatible. With NooJ, a CFG is a RG that contains recursive calls; a CSG is a CFG that contains variables and contextual constraints; an UG is a CSG that uses its outputs to perform transformation operations on its input.

Typically, a NooJ analysis consists in sequentially applying a set of linguistic resources to a text in a bottom-up approach: dictionaries (represented by acyclic Finite-State Automata in which terminal states point to the recognized word's analysis), morphological and local syntactic grammars (Recursive Finite-State Transducers), structural syntactic and/or semantic grammars (Enhanced Recursive Transducers to handle CSGs and UGs). The most ambitious NooJ applications combine grammars from of all four types to perform transformational analyses (e.g. to compute "It is not Lea who is loved by Joe" from "Joe loves Lea [Passive] [Cleft1] [Negation]"),[9] or to perform automatic text translation from one language to another.[10]

Although NooJ's approach has been proven to be useful for a number of applications (mostly, corpus linguistics where the size of a 'large' corpus is a hundred

---

[1] The time it takes to parse a text is proportional to its length n, i.e. $O(n)$.

[2] See for instance (Kasami 1965).

[3] See for instance XLFG which is a parser for the LFG formalism.

[4] i.e. one cannot even predict if a Turing machine will parse any text in finite time. Most linguists doubt that we would need the power of a Turing machine to describe real world natural languages. (Silberztein 2016a) argued that the typical examples of phenomena that would require unrestricted grammars are "extra-linguistic" in nature (e.g. anaphora resolution).

[5] See (Linden et al. 2010), (Schmid 2005) and (Karttunen et al. 1997).

[6] See (Gazdar 1988).

[7] See (Kaplan Bresnan 1982) and (Dalrymple 1995).

[8] See (Silberztein 2016a). NooJ is a free, open-source linguistic development environment available at www.nooj-association.org and supported and distributed by the European Metashare platform.

[9] (Silberztein 2016b) shows how NooJ produces several millions of transformational variants for the simple sentence "Joe loves Lea".

[10] Translations are performed just like transformations; the only difference being that the translated lexemes are obtained via a lookup of a multilingual dictionary.

megabytes at most), it does not scale up well, and NooJ cannot process large amount of texts to search the WEB for instance, or filter out all tweets in real time (100,000 per minute).

From a computational point of view, the different types of machines (.nod files for dictionaries, .nof for inflection, .nom for morphology, .nog for syntax) have imposed different parsers and a complex architecture, thus making it impossible to combine machines and thus to perform some global optimization.

2   The RA Approach

The basic principle behind the new RA linguistic engine is to unify all machines, so that only one parser will be used to perform all analyses. Thereafter, it will be possible to merge and compose them instead of applying them in sequence. For that to happen, we need to solve a series of problems, among them the following:

– how to represent morphological rules so that they can be used both to generate forms and to lemmatize them;
– how to represent CFGs and CSGs with finite-state machines.

2.1   Reversible Morphological Grammars

In NooJ, all linguistic resources are supposed to be application-neutral; in particular, they should be useable both by parsers and by generators. Although this is indeed the case at the functional level, internally NooJ's morphological parser and generator work quite differently: NooJ inflects and derives words by applying morphological recursive transducers to lemmas, but lemmatizes forms by looking up automata of a dictionary. Morphological paradigms are described by enhanced CFGs such as the following:

TABLE = <E>/singular | s/plural;

Paradigm TABLE states that if one adds the empty string (<E>) to a lemma (e.g. *pen*), the resulting word form (e.g. *pen*) is in the singular; if one adds an 's' to the lemma, the resulting word form (e.g. *pens*) is in the plural, i.e.:

pen + <E> → pen/singular
pen + s → pens/plural

From the following dictionary entry:

pen, NOUN + FLX = TABLE

An exploration of the transducer compiled from rule TABLE produces the following output:[11]

pen, pen, NOUN + FLX = TABLE + singular
pens, pen, NOUN + FLX = TABLE + plural

---

[11] The input/output result produced by the corresponding RA Finite-State Machine is underlined.

All the forms generated by morphological transducers are then stored in an acyclic finite-state automaton built using a variant of (Daciuk et al. 2000)'s linear minimization algorithm. However, the fact that the resulting automaton needs to store the original lemma (e.g. *pen*) for each of its entries degrades the efficiency of the minimization algorithm considerably.

Rather than using a transducer (the morphological grammar) to produce inflected forms, and an automaton (the dictionary) to lemmatize forms, RA uses only one machine in both directions. To lemmatize an inflected form, RA applies the machine compiled from grammar TABLE "in reverse":

Pen – <E> → pen
Pens – s → pen

Since we can now compute the lemma from each of its inflected forms, it is no longer necessary to store the lemma in the dictionary: therefore all the forms that share the same analysis (e.g. *beds*, *cars*, *engines*, etc.) will be associated with one unique terminal state. However, it is not always possible to simply "reverse" a morphological grammar. For instance, consider the following paradigm:

MAN = <E>/singular | <B><B>en/plural;

The <B> operator (for "Backspace") deletes the letter located on top of the word stack (i.e. at its end). This paradigm produces the two forms *man* and *men* from the lexical entry *man*:

man + <E> → man/singular
man + <B><B>en → men/plural

If we simply reverse it, we get:

man – <E> → man
men – ne<B><B> → m??

The generation process used the <B> operator to delete a letter, but the lemmatization process doesn't know which letter was deleted in the first place. More generally, NooJ morphological grammars are not reversible.

The new RA framework thus proposes to replace the former NooJ grammar with the following one:

MAN = <E>/s | <Bn><Ba>en;

i.e. one adds to the former destructive NooJ operators the information required to reverse them. The new grammar can then be used both to generate inflected forms and to lemmatize them:

man – <E> → man
men – ne<Ba><Bn> → man

Just like all other linguistic resources, RA's morphological grammars can be used both to generate all the inflected forms from a given lemma, and to lemmatize every inflected form.

Finally, we note that during the process of inflecting a lemma using initial NooJ's operator <B>, one knows exactly what letter is being deleted (it is on top of the stack): an automatic compiler can then use this information to compute the RA grammar equivalent to the initial NooJ grammar, just by simulating the inflection process.

2.2   Context-Free Grammars

In Context-Free Grammars, $\alpha \rightarrow \beta$ rules are such that α contains one and only one non-terminal symbol, whereas β can contain any sequence of terminal and non-terminal symbols. In NooJ, CFGs are written either in a text form:[12]

$$\alpha = \beta;$$

or graphically, by a set of recursive graphs. For instance, Fig. 1 presents a grammar that contains 3 graphs: the top graph (called Main) contains references to graph NP and graph VG.

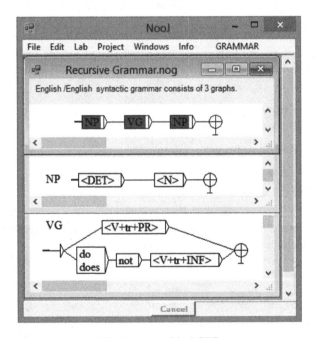

Fig. 1. A graphical CFG

The RA linguistic engine 'flattens' recursive graphs by removing left and right recursions. Middle recursions typically correspond to embedded structures such as in the following sentences:

---

[12] In NooJ CFG grammars, β is a regular expression, built on terminal and non-terminal symbols and <E> (empty string), e.g.: NP = (<DET> | <E>) <ADJ> * <NOUN>.

The cat saw the mouse
The cat (Joe likes) saw the mouse
?*The cat (the cousin (Eva talked to) likes) saw the mouse
*The cat (the cousin (the neighbor (you know) talked to) likes) saw the mouse

Although there are over 30 languages that are currently being formalized with NooJ by their native speakers, we are yet to find good examples of unlimited middle recursions that feel natural beyond two levels of embedded phrases.

The new engine RA 'cheats' by replacing all middle recursion references in a graph with a fixed number (typically two) of copies of the corresponding embedded graph. As a consequence, the last sentence of the previous example would not be recognized by RA's parser. However, the benefit is that RA can compile any NooJ CFG grammar into a finite-state machine. If the size of the initial NooJ grammar is n, the size of the corresponding RA grammar could be n^3 in the worst case; in practice though, all the NooJ grammars we encountered produced RA grammars that were lower than eight times the size of the initial NooJ grammars.

2.3   Context-Sensitive Grammars

In Context-Sensitive Grammars, production rules are of the type $\gamma A \rightarrow \gamma\delta$, where A is a non-terminal symbol, and δ and γ are sequences of non-terminal and terminal symbols (γ describes the context in which A can be rewritten into δ).[13] These rules can be rewritten in NooJ by adding the following rules:

$R_i \rightarrow R_i$ ($context γ) A
A $\rightarrow \delta$/<$context>

Where rules R_i are added for each rule in the original grammar; these rules are used to define a $context variable that will be set every time context γ appears before A. The second rule uses constraint <$context> to check that variable $context has indeed been defined (if not, the rule is rejected and A is not rewritten as δ).

Although this translation shows that variables and constraints give NooJ the power to process any CSG, using variables and constraints is much more natural in practice. (Seljan et al. 2002) gives several examples of typical formal CSGs, e.g. $\{a^n b^n c^n d^n, n > 0\}$, $\{a^n b^m c^n d^m, n > 0\}$, etc. (Silberztein 2016a) shows how to translate them into NooJ graphs that are much easier to understand. For instance, the following grammar recognizes languages such as $\{a^n b^n c^n d^n e^n, n > 0\}$:

Without taking into account the constraints shown in bold in the graph, this grammar recognizes the regular language a*b*c*d*e*, storing the sequence of a's in variable $A, the sequence of b's in variable $B, the sequence of c's in $C, the sequence

---

[13] This is the definition of left context-sensitive grammars. In right context-sensitive grammars, the non-terminal symbol of the left hand side is followed by the context, i.e. production rules look like: $A\gamma \rightarrow \delta\gamma$. The equivalence of left and right context-sensitive grammars was established by (Penttonen 1974). Another, more general definition is that context-sensitive grammars contain rules such as $\gamma A \gamma' \rightarrow \gamma\delta\gamma'$. (Kuroda 1964) proves that all these grammars have the same power of description.

of d's in \$D and the sequence of e's in \$E. Thereafter, the parser checks that properties \$LENGTH for the four variables are identical. Note that "mildly context-sensitive formalisms" such as CCG or TAG formalisms can represent languages such as $\{a^n b^n c^n, n > 0\}$ but not the one in Fig. 2.

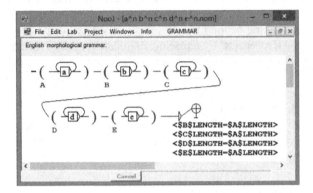

Fig. 2. $\{a^n b^n c^n d^n e^n, n > 0\}$

Figure 3 is an example close to linguists' needs:

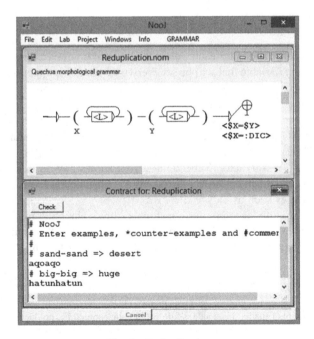

Fig. 3. Reduplication

<L> matches any letter. This morphological grammar recognizes word forms that contain the same affix twice (<$X=$Y>), and also that the affix is a valid lexical entry (<$X=:DIC>). For instance, in Quetchua, the word "ago" [sand] can be derived in "agoago" [sand-sand = desert]. Many languages use reduplications, such as Indonesian, Tagalog, Japanese, Mandarine, Quechua, etc.

As for CFGs, RA can compile a NooJ CSG into a corresponding deterministic finite-state machine. RA parses a text of length t by applying the finite-state machine in two steps:

- During the first step, RA ignores all constraints; this parsing can be performed in $O(t)$.
- The parsing process may produce multiple intermediary solutions;[14] the number of solutions is $O(t)$.
- The RA engine needs to validate each solution by checking the corresponding constraints.

NooJ constraints are of three types:

- existence (e.g. <$context>) to check if a variable has been set or is undefined;
- string equality (e.g. <$X=$Y>) to check if two lexemes are equal;
- symbol matches (e.g. <$X=:VERB>) to check if a lexeme matches an annotation and its properties.

All constraints are implemented by simple unification operations like the ones in systems such as LFG or HPSG, but they are not recursive (because Nooj's annotations are 'flat' sequences of atomic property/value pairs, rather than trees). Therefore, checking each constraint can be performed in constant time; the maximum number of constraints to check for a solution is proportional to the size of the grammar g; therefore, RA can apply CSGs to texts in $O(g\ t)$.

2.4   Compiling Syntactic Grammars

In NooJ, there is a fundamental difference between grammars that operate inside word forms at the character level, and grammars that operate at the phrase/sentence level at the lexeme level. In consequence, NooJ uses two different parsers and two different machines (.nom and .nog).

RA can process NooJ morphological and orthographical grammars with a simple format conversion; however, transforming a NooJ syntactic grammar into a RA grammar necessitates human intervention. Consider the graph of Fig. 4 below.

- All the connections to or from an empty node should be translated into ε-transitions;
- All the connections from an English word to another one (e.g. from "April" to "the") must be translated into transitions labeled with a space character " ";

---

[14] This is the case for the grammar of Fig. 3, for which the parsing process produces w-1 intermediary solutions because there are w-1 ways to split a word form into two non-empty affixes.

– Connections from an English word to some punctuation mark (e.g. "Monday" to
","") must be translated into ε-transitions; connections to other punctuation marks
(e.g. "—") require spaces.

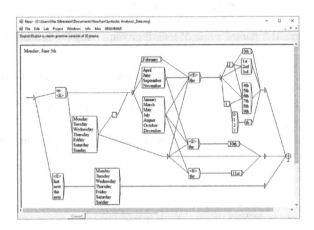

Fig. 4. A syntactic grammar

Choosing between spaces and ε-transitions is not straightforward around symbols
because a given NooJ symbol (e.g. <VERB>) can refer both to a single lexeme (e.g.
"eating") or to a sequence of word forms (e.g. "is willing to stop eating"). Lastly, NooJ
recursive grammars might include embedded graphs that recognize the empty string: to
translate a NooJ syntactic grammar into a pure RA grammar (at the character level)
involves computing all the *Follow Sets* of a grammar, i.e. the sets of prefixes that are
recognized after each embedded graph.

Note finally that these translation rules depend on the text language; for instance,
the colon character must follow a space in French, but not in English; contracted words
such as *cannot* for <can> <not>) are exceptions that must be taken into account;
agglutinative languages such as Arabic might contain sequences of words without
spaces, etc.

2.5   Compositions and Optimizations

Because all NooJ linguistic resources are now compiled into a single type of finite-state
machines, it is possible to merge and compose them. There are several remaining issues
though:

– Priorities: each NooJ dictionary is associated with a level of priority that allows
users to impose or filter out lexical solutions. Because this priority system is known
in advance (i.e. before parsing takes place), RA's compiler can produce a single
dictionary that merges all dictionaries, storing only higher-priority entries into
account;

– Symbols: a typical NooJ grammar will contain lexical symbols such as <eat> (to represent all the morphological variants of *to eat*) and syntactic symbols such as <ADV> (to represent all Adverbs). These symbols must be matched by applying the corresponding RA machines. Therefore, before applying a given grammar to a text, RA must first compile the machines for each symbol that occurs in the grammar. This operation is equivalent to composing the finite-state transducer of a given grammar with the finite-state transducers that correspond to each of its symbols.

3 Conclusion

In this paper, we have shown that it is possible to build a system that allows linguists to construct regular, context-free and context-sensitive grammars in a simple and unified way, and yet process these grammars using finite-state machines. The idea is to "flatten" Context-Free Grammars, and to use variables and constraints after the matching process to implement Context-Sensitive Grammars.

Having one unified machine for these three types of grammars allows us to compose and/or merge regular grammars (useful to describe morphology) with context-free grammars (useful to describe the structure of sentences) and context-sensitive grammars (useful to describe agreements and contextual constraints).

Moreover, this framework allows us to translate a complex series of cascading transducers into one single transducer. Although computing the final transducer is costly (exponential when flattening a recursive grammar), the resulting machine can then be applied to texts in $O(g\ t)$, which is ideal for industrial applications.

References

Chomsky, N.: Three models for description of language. In: IEEE (IRE) Transactions on Information Theory IT-2, pp. 113–124 (1956). Reprinted in Readings in Mathematical Psychology, vol. 2, pp. 105–124. Wiley, New York (1965)

Daciuk, J., Mihov, S., Watson, B.W., Watson, R.E.: Incremental construction of minimal acyclic finite-state automata. Comput. Linguist. **26**(1), 3–16 (2000)

Dalrymple, M., Kaplan, R., Maxwell, J., et al.: Formal Issues in Lexical-Functional Grammar. CSLI Publications, Stanford (1995)

Gazdar, G.: Applicability of indexed grammars to natural languages. In: Reyle, U., Rohrer, C. (eds.) Natural Language Parsing and Linguistic Theories. Studies in Linguistics and Philosophy, vol. 35, pp. 69–94. D. Reidel Publishing Company, Dordrecht (1988)

Kaplan, R., Bresnan, J.: Lexical-functional grammar: a formal system for grammatical representation. In: Bresnan, J. (ed.) The Mental Representation of Grammatical Relations, pp. 173–281. MIT Press, Cambridge (1982)

Kasami, T.: An efficient recognition and syntax-analysis algorithm for context-free languages. Technical report, AFCRL-65–758 (1965)

Linden, K., Silfverberg, M., Pirinen, T.: HFST tools for morphology: an efficient open-source package for construction of morphological analysers. University of Helsinki, Finland (2010)

Seljan, S., Vučković, K., Dovedan, Z.: Sentence representation in context-sensitive grammars. In: Suvremena lingvistika, vol. 53–54, pp. 205–218. Hrvatsko filološko društvo (2002)

Kuroda, S.-Y.: Classes of languages and linear-bounded automata. Inf. Control **7**(2), 207–223 (1964)

Penttonen, M.: One-sided and two sided context in formal grammars. Inf. Control **25**(4), 371–392 (1974)

Silberztein, M.: *Joe loves lea*: transformational analysis of direct transitive sentences. In: Okrut, T., Hetsevich, Y., Silberztein, M., Stanislavenka, H. (eds.) NooJ 2015. CCIS, vol. 607, pp. 55–65. Springer, Cham (2016a). https://doi.org/10.1007/978-3-319-42471-2_5

Silberztein, M.: Formalizing Natural Languages: The NooJ Approach. Wiley-ISTE, London (2016b)

Schmid, H.: A programming language for finite-state transducers. In: Proceedings of the 5th International Workshop on Finite State Methods in Natural Language Processing (FSMNLP), Helsinki, Finland (2005)

Karttunen, L., Tamás, G., Kempe, A.: Xerox finite-state tool, Technical report, Xerox Research Centre Europe (1997)

Author Index

Printed in the United States
By Bookmasters